How to Speak Dog

MASTERING THE ART OF
DOG-HUMAN
COMMUNICATION

STANLEY COREN

THE FREE PRESS
New York London Toronto Sydney Singapore

*f*P

THE FREE PRESS
A Division of Simon & Schuster Inc.
1230 Avenue of the Americas
New York, NY 10020

THE FREE PRESS and colophon are trademarks
of Simon & Schuster Inc.

Book design by Ellen R. Sasahara

Manufactured in the United States of America

3 5 7 9 10 8 6 4

Library of Congress Cataloging-in-Publication Data Is Available

ISBN 0-684-86534-3

This book is dedicated to my longtime friend
and respected colleague Peter Suedfeld,
to his wife, Phyllis Johnson,
and their non-dog,
Buckshot

Contents

Preface

Man has great power of speech, but the greater part thereof is empty and deceitful. The animals have little, but that little is useful and true; and better is a small and certain thing than a great falsehood.

—LEONARDO DA VINCI,
Notebook, circa 1500

There is a traditional story which tells us that King Solomon had a silver ring that bore his seal and the true name of God. This ring gave him the ability to understand and to speak with animals. When Solomon died, that ring was hidden in "a great house with many doors." When I was young, I wished that I had a ring so that I could speak with my dogs.

While I knew that this was just a folk tale, as an adult I became more inclined to believe that wise King Solomon could speak with animals, even without the magic ring the legend tells of, because you and I can also learn to do this. The "magic" in Solomon's ring is an understanding of how animals communicate, and it is hidden in *science,* which is the house of many doors. The knowledge that is needed is similar to the knowledge you need to speak any language. To speak to a dog, you must first learn the vocabulary—specifically, what constitutes the "words" in his canine language. You must also learn the "grammar" of the language, namely, how to string those words together and combine them, so that you can form "sentences" that can be used to send and receive meaningful messages.

This book is about dog communication: how they "speak" to each other, how they understand messages that humans send to them, and how humans can translate the ideas the dog is trying to transmit. Understanding how dogs communicate allows a much greater ability to know what they are feeling, what they are thinking, and what their intentions are. It also provides more ways to tell them what you want them to do and to control their behavior. This doesn't mean that one can have profound conversations with dogs about natural history, moral philosophy, or even the latest Hollywood films. On the other hand, I find that my conversations with dogs are often richer and more complex than those I have with my two- and three-year-old grandchildren, and often they deal with very similar topics. Knowing canine language also prevents some common misunderstandings between human beings and canines.

During our "language lessons" we will learn about some remarkable dogs, and just how clever some everyday canines can be. We will also see how humans have affected the linguistic ability of dogs during the long history of the domestication of our first animal companions.

Some of my scientific colleagues may object to the use of the word "language" when I speak about the communication ability of dogs. It has long been believed that language is an ability confined to humans. It will become clear, however, that there is a great deal of similarity in the communication patterns of humans and dogs. As a psychologist, I am perfectly happy drawing conclusions about human learning based upon data obtained from rats or monkeys, and the same is true of most researchers. This would be clear folly if one believed that human learning is radically different in kind from that found in other animals. Therefore, I'm often surprised that when the question turns to language and communication, some behaviorists lose their belief in the continuity of abilities across species and insist that there is a radical difference separating human language from animal communication. Whether "true" language is unique to humans is an interesting question with a long and fascinating history that will unfold during our attempt to learn how to understand and speak the language of dogs.

I would like to thank my wife Joan, who made extensive comments on the first draft of this manuscript, and also our daughter Karen, who had some helpful suggestions as well. I would also like to thank my dogs, Wiz, Odin, and Dancer, for their subtle explanations of certain aspects of canine language.

How to Speak Dog

Conversations with Canines

The argument was very sound,
And coming from a master's mouth
Would have been lauded for its truth.
But since the author was a hound,
Its merit went unrecognized.

—JEAN DE LA FONTAINE (1621–1695)
"The Farmer, the Dog, and the Fox"

It is probably the case that virtually every human being has, at one time or another, wanted to be Dr. Dolittle, or to own King Solomon's ring, so that he or she could understand and talk with animals. For me, the animals that I most wanted to speak to were dogs. I remember one Sunday evening, I was sitting on the living-room floor in front of the big family radio with my beagle, Skippy. I was leaning against the side of an overstuffed chair waiting for a regularly scheduled radio show featuring my favorite movie star. The theme music started—I think it was actually the folk tune "Green Sleeves"—and then a few moments later I could hear her voice. She was barking in the distance and coming closer every second . . .

Long before our current wave of canine movie stars, such as Benji and Beethoven, and their television counterparts, Eddie, Wishbone, and the Littlest Hobo, there was Lassie. She was much more than a dog; she was a friend and devoted companion. She was a guardian of the right, a courageous protector, and a fearless fighter.

The dog that may have done the most to shape the popular conception of dogs and their intelligence was a character born in a short story published in the *Saturday Evening Post* by Eric Knight in 1938. The story was so well received that Knight later expanded it into a best-selling book in 1940, and in 1943, it was translated into a heartwarming tearjerker of a movie called *Lassie Come Home*. It was filmed in rich colors and set in Britain, where Lassie's poor family is forced by their financial troubles to sell their faithful collie to a wealthy dog fancier (whose daughter is played by a very young Elizabeth Taylor). Lassie escapes from the Duke of Rudling's harsh kennel keeper and manages to work her way from Scotland to England to get home to her young master (who is played by Roddy McDowall). The role of Lassie was not portrayed by a lovely female dog at all, but by a male dog named Pal. In fact, all of the Lassies ever since have been female impersonators. Male collies were preferred to play the part, since they are larger and less timid than female collies. More important, when an unspayed female dog goes into heat (which they do twice a year), she often loses much of her coat. It would be very distressing to movie watchers, and it would be a film editor's nightmare, to have the fullness of Lassie's coat vary from one scene to another.

Gender issues aside, Lassie had a huge impact on our concept of how dogs think and act. This was partly due to the volume of material about her that we were exposed to. So far there have been ten feature films showing her exploits. In these Lassie managed to upstage some of the greatest stars in Hollywood, including James Stewart, Helen Slater, Nigel Bruce, Elsa Lanchester, Frederic Forrest, Mickey Rooney, and many others. There was also a TV show which ran from 1954 through to 1991 (with a few interruptions), using six different settings and rotations of cast. At times, Lassie's families included such familiar actors as Cloris Leachman and June Lockhart. Many of these episodes are still appearing on television in syndicated reruns today. There was even a Lassie cartoon series (Lassie's Rescue Rangers) that played on Saturday morning TV for the kids.

Perhaps Lassie's most unusual starring role was in a radio series, which ran from 1947 through 1950, and I was one of her young fans. I'll bet that given the media mentality of today, producers of a radio series involving a dog might argue that it was necessary to give Lassie a human voice, so that we could hear her thoughts and know what she wanted to say. It would be a soft female voice, of indeterminate age, perhaps with a slight Scottish accent to remind us of her origin. These early radio episodes, however, were true to the character of Lassie on the screen. She never spoke human language, she barked. It is interesting to note that Pal actually did the barking on the radio show; however, the whining, panting, snarling, and growling were all convincingly done by human actors.

One part of the magic of the show was that Lassie did not have to speak in English, Spanish, German, French, or any other human tongue. Her family and everybody who heard her understood her completely. An episode might typically go like this:

Lassie runs out into the field, barking and whimpering frantically. Her young master asks, "What's wrong, girl?" and Lassie barks.

"There's something wrong with Mom?" he interprets, and Lassie barks and whimpers.

"Oh no—she's hurt herself! Dad told her not to use that machine by herself. You go get Dr. Williams. I saw him stopping by the Johnson place just a little bit down the road. I'll go back to see if I can help."

The boy runs across the field toward home. Lassie barks and races off for help. The doctor will, of course, understand every bark and whine and come to the rescue as well.

In other episodes and at other times, Lassie's barks tell of bad men coming, of hidden or stolen goods, or alerts her master that someone is either lying or speaking the truth. It seems that Lassie speaks a universal speech. There is one episode with a boy from France, who comes to live with his uncle after his family dies tragically. This poor child speaks no English. Fortunately, he doesn't have to. Lassie speaks the universal language of dogs (let's call it "Doggish"). He, of course, understands it immediately, since apparently all French dogs use the same language. Because of this, Lassie is able to tell him (with more barks, whimpers, whines, and an occasional muted growl) that he has come to a place where people want to be his friends, although there is one bad boy he should watch out for. Lassie comforts him, integrates him into the community, settles some misunderstandings between him and the local children, and then teaches him his first few words of English, which are, of course, "Lassie, you are a good dog!"

I really felt jealous of Lassie's family and neighbors. They could all understand the language of dogs, and they knew how to make their own dog understand exactly what they were saying as well. I fondled Skippy's long, leathery ears and wondered why I was so linguistically inept.

It's not that I couldn't understand anything that Skippy was trying to tell me. When his tail wagged, I knew that he was happy. When his tail was tucked under his belly, I knew that he was feeling poorly. When he barked, I knew that someone was coming, or that he wanted to eat, or that he wanted to play, or that he was excited . . . Well, he barked a lot. When he bayed (that little yodeling sound that beagles make), I knew that he was happily tracking something. The linguistic failings were not Skippy's, they were mine. Sometimes my dog would be incredibly innovative in telling me what he wanted. There was the day he deliberately pushed his water dish across the kitchen floor until it banged against my shoe, just to tell me that he was thirsty and the bowl was empty. Still, most of the time I just couldn't understand what he was saying and our lack of communication made me very sad. Now, after many years of re-search and study, I think I am beginning to understand the language of my canine friends. As a psychologist, I have also come to realize how an understanding of dog communication can affect human-dog relation-ships.

In humans, language often appears to be the single most important element in determining successful social relationships and general ad-justment. When you look at the research on the relationship between children with disabilities and their families, you find that love and affec-tion can be fostered and maintained even though the child suffers from massive problems, as long as the child can speak and understand lan-guage at a useful level. The families of children who have many fewer difficulties, but whose language ability is impaired, report more severe social and adjustment problems, and seem to feel less affection and more frustration with the child. Similarly, several studies have shown that the single most important factor in determining whether an immigrant or refugee will integrate well into their new society is the speed and profi-ciency with which they learn the language of their new country. In much the same way, a human's ability to understand the language of the dog can determine how well the dog is accepted into the family.

Misreading a dog's emotional state can be distressing for its human family, and can even be fatal for the dog. Consider the case of Finnigan, a beautiful Irish setter from a kennel run by a woman named Melanie. I knew Melanie as a careful breeder, whose conscientiousness had allowed her to create a line of dogs that were not only physically handsome but also warm, playful, and tolerant. With that in mind, you can imagine Melanie's distress when she received a phone call from the family that had bought Finnigan. They complained that he was too aggressive. They said he was leaping and snarling at visitors and other dogs. When these problems arose, the family had called in a trainer, but he had found the dog difficult to handle and failed to eliminate these aggressive displays. In the end, he had recommended that the dog be euthanized. The family didn't want to do this, but felt they couldn't keep him under the circum-stances. Melanie offered a refund of the purchase price and asked that the dog be sent back to her.

Then she called me up. "I've never really had to deal with an aggres-sive dog before," she said. "I was wondering if you could be with me when I go to pick him up—just in case there's something I can't handle."

I couldn't imagine one of her dogs being aggressive, but the worry in her voice was such that I agreed. I was there to help pick Finnigan up. I had brought with me the usual accoutrements for dealing with virtually any type of aggressive dog. There were a couple of strong leashes, a slip collar, a head halter, a muzzle, and even a large heavy blanket in case the dog had to be physically restrained by wrapping him so that some of this control equipment could be applied. In addition, I brought a pair of heavy leather gloves (which have literally saved my skin a few times).

When the truck carrying Finnigan arrived, I bent down to look into the front of his tan plastic carrying kennel. No snarls, no growls, just an excited whimper. Still, caution seemed like the best plan, so we opened the door slowly. Out leapt this happy red dog, who looked around, trying to discern where he was. Then, in a response that was clearly triggered by the unfamiliar surroundings of the loading dock, he showed every tooth in that large mouth of his.

My response was involuntary, but I think that I upset Melanie when I began to laugh. I realize that to a person who doesn't understand the language of dogs, this flash of forty-two long white teeth could easily have been interpreted as an aggressive display. However, there are different ways that a dog can show its teeth, and the expression Finnigan was wearing was actually a submissive and pacifying grin. This expression did not mean, "Back off or I'll bite," but rather, "It's okay. I'm not a threat. I understand that you're the boss around here."

The young setter's bounciness did cause him to leap at people and other dogs. But this leaping was done as part of a greeting. He simply wanted to touch noses with those tall two-footed dogs that we call humans, and the only way to reach their nose was to jump up. To ensure that this would not be viewed as a threat, he did it with a submissive grimace. The more he was corrected by the family and trainers for his "aggression," the more submissive he became. The more submissive he was feeling, the wider he "smiled," reasoning that they had simply missed his signal and he truly wanted to pacify the situation. Of course, the wider he "smiled," the more teeth he showed.

Finnigan's first family simply didn't understand what the dog was trying to say; had they followed the advice they were given, they might have put this handsome red dog into an early grave. Finnigan now lives happily with a new family. Melanie tells me that he still smiles and jumps a bit, but she has explained what this means to his new masters. Because they understand his message, they know that he is safe to love.

Unfortunately, mistranslation of the signals that a dog is giving is quite common and can lead to serious problems and bad feelings. A woman named Eleanor came to me with a problem. It involved Weedels, a blond American cocker spaniel, who, according to her mistress, was "driving my husband crazy. She simply refuses to be housebroken, and is now making puddles simply out of spite. Stephen [her husband] says if we can't solve this problem quickly, we'll have to get rid of her."

The period of time while a puppy is learning to be clean in the house is often stressful. It is usually solved within a few weeks, however, if care is taken to regulate the dog's food and water intake, and the owner is alert to the times when the dog should be taken out to empty its bladder and bowels. In this case, Weedels was nearly seven months old, which seemed a bit old not to be housebroken. So I asked what they had done to train her.

"Stephen likes things in the house to be neat and clean, so it was important that we housebreak Weedels early. I read one of those books on puppies and followed its advice, and we got her to make her stool outside. But we still occasionally had 'wet accidents.' Stephen said I was being too easy on Weedels and he would solve the problem. When he found a place where she'd wet the floor, he dragged her over and rubbed her nose in it. Then he yelled at her and gave her a slap on the rear when he put her outside.

"Stephen went away on a sales trip and was gone for nearly four weeks. During that time, Weedels was fine. Maybe there were one or two accidents, but that was all, and I just cleaned it up and put her out in the yard without a whole lot of fuss. The last two weeks, things were absolutely clean. Then, just a few days ago, Stephen came back and everything fell apart. You wouldn't believe what this dog did. The moment Stephen walked into the house, she peed on the floor right in front of him. He got so angry I thought he was really going to hurt her. Weedels just seems to want to annoy him. Whenever Stephen walks into the room, she crouches way down low and makes a puddle in front of him just for spite. Yesterday was the last straw. Stephen walked into the room and Weedels rolled on her back, like dogs sometimes do for a belly rub. When he bent over her, she tried to pee right into his face! That's why I'm here today."

My heart went out immediately to poor Weedels. Dogs do not communicate by using the same signals that humans do. In this case, Weedels was giving a clear message in the only language that she knew. Unfortunately, there were no translators around, so her plea for understanding was being misread and getting her into trouble. Her problem had nothing to do with housebreaking. From my conversation with Eleanor, I knew that Weedels was almost completely housebroken by now. The problem had to do with her husband, Stephen. In his early interactions with Weedels over her urinating on the floor, he was particularly harsh in his corrections. This caused Weedels to become quite fearful of him. If a dog is experiencing a large amount of social fear, it will try to make itself appear to be as small, insignificant, and non-threatening as possible. Crouching low to the ground or rolling over onto its back are part of this pattern.

What Eleanor thought was a spiteful attempt to urinate on her husband's face was simply the release of urine from a dog who was rolling into a very submissive and frightened position. The urine was designed to remind the "dominant dog" of puppy behaviors. Puppies need to be cleaned of urine and feces when they are small, and the mother usually simply rolls them on their backs to do this. Thus, Weedels was really trying her best to say, "You frighten me, but look, I'm no threat. I'm nothing more significant than a helpless puppy." Once Weedel's message was translated for Eleanor, the situation became much clearer. Now her task was to try to build Weedels's confidence. A larger problem might be trying to get her husband to be gentler and less threatening around the dog.

Many common canine messages can be misinterpreted. A woman named Josephine once asked me to help her with a problem she was having with her dog.

"Bluto is acting far too affectionately toward me and it bothers me and upsets my husband. He originally got Bluto as a guard dog and he doesn't want him acting like a wimp, even around the family," she told me over the phone.

Bluto turned out to be a large, dark Rottweiler, who had been named after the big, bad, ugly cartoon character who was always fighting with Popeye the Sailorman. The name, which had been given by Josephine's husband, Vincent, told me something about the man and his expectations for the dog. Vincent was a forceful trainer, and had often used fairly harsh methods to enforce his will over Bluto. The dog obeyed him, although sometimes with apparent reluctance. According to Josephine, Bluto didn't obey her at all, but did show extreme and persistent signs of affection toward her.

When I arrived at their home, Vincent was at work and Josephine brought me into her living room. I sat on a chair and looked across at her, sitting primly near one end of the sofa with Bluto beside her on the floor. Bluto appeared to be around 120 pounds of hard muscle, while Josephine appeared to be around 100 pounds, very slight, and not particularly athletic. As we talked, Bluto placed his paw on her knee, and she immediately responded by stroking his head. After a few moments, Bluto jumped up onto the sofa beside her, and Josephine moved slightly to one side to accommodate his great bulk. He sat there, looking occasionally at me, and then staring at her. When he looked directly into her eyes, she would reach her hand up and lightly stroke the side of his face.

Next, Bluto leaned his weight against the small woman. After a few moments, she shifted to the side to be free of the pressure of the heavy dog. He reacted by shifting his position so that he was again sitting beside her and once more leaning against her. Again, she moved away a few inches and again the dog moved closer. As we spoke, this spectacle continued until Josephine had been forced all the way to the far end of the sofa. At the point where she could no longer move any farther, she stood up in exasperation and pointed at the dog.

"This is exactly what I mean. He's always asking for attention by pawing at me. He's always staring into my eyes and leaning against me to show me how much he loves me. I can't even watch a television show without him pushing me off the sofa unless Vince is here. I don't want to hurt his feelings, but he's a big dog. That kind of continuous affection from such a large animal is annoying and disturbs my husband. Is there some way we could train him to be less dependent and more confident and independent?"

Once again, a message had been sent by a dog and misinterpreted by the human receiver. Bluto was not telling Josephine, "I love you. I need you. I'm totally dependent upon your affection," which was the translation she and her husband were giving to these signals. Instead, Bluto was saying, "I am higher status than you are. When the leader of the pack [Vincent] is away, then I'm in charge, and you will give way to me and respond to my needs."

The signs of dominance were all quite readable. A dog who puts his paw on a human's knee is often expressing dominance over that human, in the same way that a wolf will put his paw or head over the shoulder of another wolf to demonstrate that he is of higher status. Bluto's staring directly into Josephine's eyes is a classic dominance and threat gesture, designed to produce pacifying responses in other pack members. Josephine was accepting his dominance by stroking the side of his face, in the same way that a wolf of lower dominance might lick the face of a higher-status dog. Finally, his leaning behavior was designed to make the small woman give way. Pack leaders can occupy any part of the territory they desire and can sit or sleep where they want to. Lower-ranking members of the pack move away to permit this, thus accepting the other animal's dominance. In other words, everything Bluto was "saying" pointed to "I'm boss," and everything that Josephine was "saying" was "Yes, I humbly accept your authority."

Once the message became clear, the solution to the problem could easily be found. In the end, Josephine had to take Bluto to basic dog obedience classes, where he learned to follow her commands. Since she couldn't physically dominate the dog, she used treats to induce him to respond. She also became completely responsible for his feeding at home, and required him to respond to simple commands like "Sit" and "Stay," before he would be fed. In the wild, it is the leader of the pack who eats first and controls the hunt and distribution of food. By controlling food, in the form of meals and treats, and insisting upon Bluto's responding to her commands to get these, Josephine was now using a form of dog communication to tell him: "This two-footed dog is of higher status than you, even if I'm not as large or strong as you are."

In the same way that people can learn to interpret the language of dogs, there is no doubt that we humans can communicate with dogs if the person chooses to speak in their canine language. An interesting instance of this was described to me by Dr. Michael Fox, who has made his mark as one of the foremost researchers on dog and wild canine behavior. At that time, Fox was a faculty member in the Psychology Department of Washington University in St. Louis. He was doing some marvelous work comparing the behavior patterns of various wild canines such as the wolf, fox, and coyote with the behavior patterns of domestic dogs. This was the work which eventually convinced scientists that there is a universal core of behaviors common for all canines. To the extent that this is true, we can learn about our pet dog by studying the behaviors of wild wolves. Conversely, we can learn about wolves by the study of a little spaniel who might be nestled at our feet. This is a well-accepted concept today, but at the time it was still controversial.

I met Dr. Fox after a lecture he had given. When I introduced myself, I mentioned that I had seen the television documentary he had worked on called *The Wolf Man*. He responded immediately and took the discussion off in an unexpected direction.

"Ahhh, yes. You know that project taught me that I did know how to communicate with wolves well enough to save myself from harm, but that I didn't know enough about what the wolves were saying to avoid difficulties in the first place."

There was a tone of amusement rippling through his mildly English-accented voice. "You see, we'd just introduced some wolves to each other in the research compound, and we were hoping we could photograph their behavior. I believed we had an opportunity to get some good film of their greeting patterns and how they sorted out their dominance relationships. In any event, the oldest male and his mate (both around

four years old) were down at the end of the research area with the rest
of the group. It turns out that the female was in heat and was doing a lot
of submissive nudging at the male. With strange wolves around in his
territory, and a mate in heat, I suppose the male was getting pretty edgy
about the whole thing.

"We'd been concealed behind some bushes, when this pair broke
away from the others and came toward the shrubbery where we were
hiding. When they passed us, I thought we could get some good pictures
of them, so I rushed after them. Suddenly they reversed their direction,
and I was caught in the act. Here was this human being rushing directly
toward them and staring at them. At the best of times, that kind of activity [running directly forward and direct eye contact] signals a threat,
so I immediately stopped moving. I thought that that would be enough
to avoid any problem. But I must have still been staring directly at them
in what had to be interpreted as a wide-eyed challenge. There were no
further words between us—the male simply attacked me.

"With a camera strapped to each of my wrists, there wasn't a lot that
I could do, so I raised my hands in the air and shouted for the handler.
[In retrospect, this was the wrong thing to do, since the elevated hands
look like another attempt at asserting dominance. Doubtless this was
read as equivalent to another animal rearing up to make itself seem
larger. The shouting might also be misinterpreted as a bark or growl-
bark.] Meanwhile, the male was biting my hand and arm and back, and
the female had joined in and was attacking my legs. It was at that point
that I finally had the presence of mind to remember how to tell them that
this attack wasn't necessary. I froze in place and huddled down to make
myself small—all the while making whines and whimpering sounds, like
a frightened and submissive wolf cub. Although they immediately broke
off the attack, the male came right in front of my face, gazing directly
into my eyes and snarling. I responded by averting my eyes and avoiding
any eye contact while still continuing to whine. When the pair seemed to
have eased off a bit, I tried to back away from them a little, but that only
made them attack again. This time, though, the attack was all threats
and no actual biting, which meant that the most important part of my
message had gotten through.

"Around that time, the handler arrived, got hold of the male, and
dragged him away. The female stayed with her eyes staring directly at
me, as if waiting for me to make the next move. I didn't. I just stayed
there, eyes half closed in submission and whimpering, until they finally
got a collar on her and pulled her away.

"Fortunately, I was wearing some pretty thick clothing, so their teeth didn't break a lot of skin. On the other hand, the pressure and the shaking when they bit me caused a lot of pain and bruising, and also some muscle and tendon damage."

He laughed lightly and sipped at his drink. "One of the people who was there took some photos of the whole incident. One of those pictures shows a perfect example of a fear grimace—only it's being displayed by a human psychologist, and not a frightened wolf."

In this case, an extremely knowledgeable and intelligent human being inadvertently gave the wrong set of signals to a canine, and this produced an attack. Fortunately for him, he knew enough about canine language to be able to convey that it was all a mistake, and that he had no intention of continuing to challenge or suggesting any further threat. This probably saved him from considerable harm.

In many respects, our ability to live successfully and happily with any dog may depend upon our competence in reading the dog's language. If a person knows how to speak "Doggish," then they can interpret what the dog is trying to say and also give unambiguous signals which the dog can translate. Unlike human languages, which have to be learned, much of the dog's language is encoded in its genes. He does have the capacity to learn to understand a lot of human language as well, which will ultimately make communication with people easier. However, before we can discuss speaking with our canine companions, it will be useful to know something about language itself.

2

Evolution and Animal Language

Before we can speak of translating the language of dogs, we should first answer a critical basic question: Do any animals, other than humans, have their own languages? Although most scientists will agree that animals can *communicate* with each other, the problem seems to lie in what we are going to define as "language." Many researchers, particularly linguists, will grant that animals can use sounds as part of their communication system; however, they go on to claim that animals don't even have the basic language elements that we can call words. According to their analysis, animals have no ability to "name" objects in the environment, such as "ball" or "tree," or to express abstract notions, such as "love" or "truth."

Noam Chomsky, the well-known linguist from the Massachusetts Institute of Technology, has offered a theory that only humans are capable of learning languages because only humans have the brain structures needed. Humans learn vocabulary at a fantastic rate. Between the ages of two and seventeen, the average child will increase its vocabulary at a rate equivalent to learning a new word every ninety minutes of its waking life. At the same time, it will acquire a complex grammar and syntax. What is most amazing is that all of this is learned without the need of formal schooling and instruction. According to Chomsky, this remarkable achievement can only be explained by assuming that all human brains come with a built-in language-processing organ. This special organ does not contain a specific language, but rather the blueprint for learning all languages. It also includes the basic structure of what he calls "Universal Grammar." This is what allows children to learn language so quickly—in effect, they already know how languages are structured because their genes have provided the information of what are acceptable or unacceptable language constructions.

My difficulty with Chomsky's theory, that language is an exclusively human ability, comes from an evolutionary viewpoint. Language clearly gives humans a great survival advantage. We can transmit or receive vital information about the state of the world and our local environment using language. We can also communicate information about past events and even about our predictions of what will happen in the future through the use of language. It makes survival a lot easier when individuals can tell each other where there is food and water, where a lion was last seen lurking about, or warn of the approach of a forest fire. Language can also be used to coordinate social interactions with other members of the group, whether individuals are organizing a hunt, arranging care for their babies, establishing social contact with potential mates, or resolving a difficulty with another individual or band in order to avoid physical conflict. Any animal species that had language would thus have a powerful tool which would make it a more successful beast in this hostile world.

Whenever there is some successful evolutionary adaptation or change in an animal, it is almost always preceded by some simpler versions. Consider the marvelous mechanical device that has given human beings the ability to create our technological world—our thumb. It is an *opposable* thumb, which means that we can touch it to the tips of any of our other fingers. This provides the ability to deftly manipulate small objects, and to create and use tools. This special digit first appeared in monkeys as a stubby thing, not really opposable against any fingertips. As the various primates evolved into apes, however, it grew longer, and in some other primates the thumb can oppose one or even two fingers to some degree. Thus, the human thumb shows evidence of evolving from simpler forms that predated it. Similarly, the spectacular flying ability of birds was preceded by less complex versions of this skill. At an earlier time, there were animals (e.g., Pterodactyls) which could glide through the air. This gliding was not true flight, but rather a sort of floating through the air with a limited degree of control. The flight ability of birds is simply a more advanced and complex evolution of this simpler ability to soar or glide. What has been added is the ability to take off from any surface and to change altitude at will.

Most important and useful abilities show some form of continuous change through the eons of evolution. What Chomsky and others who deny language ability in non-human animals are offering is what biologists call a "hopeful monster" theory. It is a miraculous accident in which a freak mutation, just by chance, happens to produce a radically better equipped animal: the evolutionary theorist's version of "divine intervention."

This kind of explanation makes me uncomfortable. Evolution is much like a big highway that species travel down. Changes in direction are fairly gradual, since too sharp a turn will cause the quickly traveling vehicles (or evolving species) to fly off the road into extinction. At the biological level, this idea of a highway shows up in the form of continuous slow change, with a lot of similarity between various animal species, especially at the genetic level. It might be surprising, or even disturbing, for some people to learn that recent findings in modern biochemistry suggest that humans are not so genetically unique as we might have thought. DNA analysis suggests that at the molecular and genetic levels, humans and chimpanzees are at least 98 percent identical. This degree of similarity is so close that some scientists have proposed it might be possible to perform crossbreeding and make a hybrid species. Of course, presumably moral and ethical considerations would forbid such a genetic experiment, but this possibility does illustrate how similar human beings are to other primates. Even an animal as apparently distant from humans as our pet dog is still quite similar to us. We are both mammals, and the DNA sequence codes of dogs and humans have better than 90 percent agreement with each other.

If we are so close genetically to non-human animals in all other aspects, it seems unlikely that evolution would make a sudden quantitative and qualitative jump when it comes to language ability. The more logical conclusion would be that evolution was heading toward the appearance of human-level linguistic abilities, and if we look hard enough, we will find a continuous series of stages that lead to our own form of language ability. These early language abilities will not appear full-blown, but the precursors should first make their appearance in the communication patterns of other animals—such as dogs. The logical expectation is that the "language" of dogs will be a lot simpler than the language of people, but that same logic suggests that there will be a language of dogs.

If the logic of the situation suggests that other animals should have some simpler form of language, then why do researchers, such as Chomsky, suggest that the human species is so special in linguistic ability? Actually, they are continuing a long tradition which began with philosophers and early naturalists who wanted to make the argument that human beings are unique and in a class by themselves. There is something in that line of reasoning that appeals to our ego. It makes us feel proud that we are so gifted, that all of nature is beneath us, and perhaps even that God has singled us out for some special blessing.

Human beings obviously differ from animals in many ways. Humans, for example, are the only animals that have large, pendulous

breasts, wear clothes, pierce body parts (such as ears) and insert orna-
ments into the holes, dye their hair, tattoo their bodies, paint their faces,
use money, or cook their food. Such minor differences bring us no pride;
rather, it has always been in the mental realms, such as reasoning,
morals, and language, that we wish to assert our uniqueness and our
superiority.

Probably the best known version of this argument was given by René
Descartes, who proposed that no animals other than humans had consciousness, real intelligence, or any sort of higher mental abilities. Non-
human animals were simply furry machines, very cleverly built, which
reacted to stimulation from the world in much the same way that a
machine responds when you throw a switch. The church supported
Descartes's conclusions, since if animals had true thinking ability, they
might also have souls. If they had souls, then this would raise certain
ethical issues about the treatment of animals, which the church did not
want to deal with, such as the morality of killing animals for food, deny-
ing them free will, or forcing them into service for humans. For Descartes,
to test whether an animal could think and had consciousness, you simply tested their language competence—specifically, their ability to cre-
atively produce spoken human language.

The idea that man was special was not universally believed, however.
The Greek philosopher Aristotle, the scholar St. Thomas Aquinas, and
the evolutionary biologist Charles Darwin all came to the conclusion
that humans and animals differ only quantitatively (in the degree to
which their mental abilities express themselves) rather than qualitatively
(in the actual nature of those mental processes). This would, for instance, allow a less complex species to have a language that is not as
complex as that of humans.

Of course, whether animals have language or not depends upon how
you define language. If you define language as any communications system or signaling system, then every living animal on this planet would
probably have to be credited with language. Crickets and grasshoppers
communicate their location and availability to potential mates with
sounds made by rubbing the rasplike edges of their hind legs together,
while fireflies communicate a similar message by flashes of light. Should
insects be credited with having language? The ethologist Karl von Frisch
thought so, and he won the Nobel Prize for his work in translating what
he called "the language of the bee."

The honeybee has evolved an extraordinary method of communication to assist in the survival of its hive. Specialized scouts search for food
and return with the news when they have found a supply of nectar or

pollen. They then inform the rest of the hive by means of a set of unique "dancing" movements. They circle the hive wall or floor in sort of a figure-eight pattern, waggling their abdomens as they dance. The specific patterns and speed of movements that they make, and the orientation and size of the patterns, convey information about the direction and the quality of the food they have found. Their movements also describe the distance to the food source, which might be several miles away.

In bee colonies there is even a special "house-hunting" scout, who appears to be uninterested in bringing back information about food supplies. This scout scours the area for a new location for another hive. If a colony finds itself with two queens, one is inevitably driven out. She gathers a band of loyal followers who will go with her to start the new community in the place found by the scout. The language of these scouts is so precise and accurate that the researcher observing the message was able to reach the new colony site before the bees themselves had arrived at their new home.

Although this is remarkable behavior, and most scientists would grant that honeybees have a complex communications system, the vast majority of researchers would refer to this as a "signaling system" rather than a true language. There just seems to be too little flexibility in the content and too much simplicity in the structure of the system to make scientists truly comfortable with calling it language. Bees seem to "talk" of nothing except "Where is the food?" and "Where shall we set up housekeeping?" No bee ever seems to say, "I'm feeling happy today," "I like you," "I find this job boring," or, "I would like to grow up to be queen myself someday."

What constitutes the minimum requirement for a true language is a difficult question and it is something that we will ultimately have to address. However, there are some aspects of human language that might not be required in all other languages. For instance, most people tend to confuse language with speech. Obviously, for humans, speech is the most common form of expressing ideas via language. In terms of evolution, voices are a rather late development. To produce spoken words requires a voice box, technically called the *larynx*. If you place your fingertips on your throat and speak or hum, you can feel the vibrations that are produced as air passes through the larynx and produces sounds. The larynx appears in higher land animals, including mammals and some reptiles and amphibians, as part of the windpipe that carries air to the lungs. Although a summer evening in the country can be filled with insect noises, no insect has a voice box, nor does any other invertebrate (that is, any animal without a backbone). Fish do not have a larynx either, since fish breathe using gills to extract oxygen, rather than lungs.

In order to understand how people are special in their ability to speak, we need to spend a couple of paragraphs on a bit of physiology. The larynx has several segments of firm, elastic cartilage held together by muscle and ligaments; it extends from the throat *(pharynx)* above to the windpipe *(trachea)* below. Because the mouth can be used both to take in food and to breathe, you need a special apparatus to separate these functions. This is the *epiglottis*, which is much like a box lid that swings over the opening between the throat and the voice box. When an animal swallows, the larynx is raised to press against the epiglottis and root of the tongue, closing off the windpipe so that food can be directed to the stomach and will not cause choking by getting into the airway.

Voice sounds are produced when air interacts with the vocal cords. These are two thin bands or folds of membrane lying across the upper section of the larynx in a V-shape. The tension on these folds is controlled by muscles. When you breathe normally, your vocal muscles are slack, which allows air to pass silently in and out of a wide slit. When the muscles are tighter, the vocal cords begin to vibrate. The tighter the vocal muscles contract the vocal cords, the higher the pitch or tone of the sound produced. This is much like what happens with a toy balloon. If you blow it up and release the pressure on the opening, the air rushes out silently. Slightly stretching the rubber at the opening to make a narrow slit causes the outward rushing to make sounds, which will vary in pitch as you vary the tension. The movements of the tongue and the lips will further modify the nature of the sounds, shaping, molding, and clipping them into new patterns.

The reason that it is important to know about the voice box is because humans and dogs are built differently and this limits the sounds that dogs can make in comparison to people. In dogs, there is only a slight bend in the airway as we go from the mouth to the windpipe. In human beings, because we stand vertically, there is a 90-degree bend in the airway; this leaves room for the larynx to lengthen, and for there to be additional soundmaking accessories, such as two resonating cavities instead of the dog's one. In addition, humans have room for a larger, rounder tongue in comparison to the shorter, flat tongue of dogs. Thus, a dog simply does not have the vocal apparatus and control to voluntarily and selectively produce various different speech sounds, as the "a" in *bay*, the "i" in *bit*, or the "u" in *usual*.

Another difference between dogs and humans is that, as hunters that often track their prey by its scent trail, canines have adapted their airways to make it easy to smell and breathe while they are running. This involves having the epiglottis locked into a closed position most of the time. It still allows canines to bark, yap, or howl while they are moving. In humans, this lidlike flap is open most of the time when we are talking.

Don't let your dog develop an inferiority complex because of its limited ability to make certain sounds. This is a very modern evolutionary development. It seems that similar difficulties were experienced by some of our fairly recent human ancestors, such as Neanderthal man. The evidence is that Neanderthals probably did not have speech, or only had limited speech abilities. Soft tissue such as a larynx doesn't survive well, so we have no fossilized vocal tracts from primitive humanoids. However, the psychologist Phillip Lieberman has demonstrated that if you try to insert a modern human vocal tract directly into the skeleton of a Neanderthal, it simply doesn't fit.[1] The modern larynx ends up in a weird and unlikely position inside the Neanderthal's chest; obviously, this is an impossible placement. We are left with the conclusion that Neanderthals probably lacked the more refined apparatus necessary to make complex speech sounds.

There is another aspect of human evolution which gives people an advantage over dogs in terms of speech production. Because humans walk upright, our hands are free to manipulate things so that we hunt and protect ourselves using weapons held in our hands. This means that we don't need a strong muzzle full of teeth to do these jobs. We can have shorter snouts and muzzles, and can allow our lips more flexibility to shape sounds. Our more flexible faces also give us the apparatus to produce a broader range of voice sounds than is possible for dogs.

Actually, evolutionary considerations such as these have led to a marvelously speculative theory that dogs may be responsible for the development of human spoken language. To follow this argument, we need to point out that there is some new evidence, based upon DNA analyses, which suggests that dogs may have been domesticated by humans for far longer than scientists had previously imagined. It is possible that dogs may have been domesticated as long as 100,000 years ago. Pushing back the origins of the domestic dog into that distant past has allowed a new wave of thinking about how dogs and humans co-evolved.

It is well established that the primitive humans who survived to become our forefathers formed an early relationship with dogs. Compare our success to that of the Neanderthals, who never got along with dogs, and who ultimately died out. Some evolutionary theorists have suggested that the survival of our ancestors had to do with the fact that our cooperative partnership with dogs made us more efficient hunters than Neanderthals.[2] With the dog's more acute sensory systems, finding game was easier. Dogs' exquisite sense of smell, combined with the adaptation of their airways to allow them to continue following a scent, even while running, made them proficient trackers. Finding game is clearly one of the most important tasks facing a hunting society.

Here is where the serious speculation begins. These theorists suggest that since these early humans now had dogs to do the tracking, they no longer needed the facial structures that would allow them to detect faint scents. This, then, allowed our early ancestors to evolve more flexible facial features, which were capable of shaping more complex sounds. In other words, our prehistoric association with dogs, who would do the smelling for us, gave us the ability to create speech.

The rival Neanderthal race, however, never formed a compact with dogs. This meant that they were left with less flexible facial features, since they still needed their better scenting abilities. Less flexibility means more limited voice control, which in turn would make speech far more difficult. Once early man had the ability to shape speech sounds, this permitted the development of spoken language. As we have already seen, language brings with it many advantages. It can help to organize a group, it can allow the passing on of knowledge and information, and it can provide us with a number of other survival advantages.

Think of it—if this theory is correct, then it may well be the case that human speech owes its very existence to our association with dogs!

Although dogs do not have speech, this does not necessarily mean that they don't have language. We know that many deaf people use signs rather than sounds as their form of language. Similarly, although evolution has denied the dog the facial flexibility, the voice box, and the degree of voluntary control needed to create human speech sounds, it is still possible that dogs can use other means to communicate. Perhaps these other forms of canine communication may have the richness and complexity needed to create a language.

3

A Dog Is Listening

There is one thing that people tend to forget when considering how individuals use language. Linguistic ability actually involves two important components. The first is the ability to *understand* language. This is really the most basic requirement. Second, and more complex, is the ability to *produce* language. It is possible to comprehend language but not to produce it. This is the case for individuals who are born mute or lose their ability to produce voice sounds through some form of accident or illness. Such people may understand what is being said, but they cannot produce the sounds which make up the language they are interpreting. We call their skills *receptive language ability* as opposed to *productive language ability,* which includes not only the ability to comprehend a language but also to produce it so that someone else can interpret it.

The earliest stages of human language development involve the development of receptive language. By the time it is thirteen months old, a human baby will typically understand nearly 100 words; however, its productive language is virtually nonexistent at this age. At thirteen months, most children will be producing one or two meaningful language sounds, while the brightest children may be producing five or six "words." Clearly, young children develop an understanding of language before they can speak.

The fact that language reception is easier to learn than language production was recognized by the U.S. Space Agency, NASA, when it started the first multinational cooperative space missions. When American and Russian astronauts were first brought together to work in space, each was required to speak their own native language. Thus, American astronauts spoke only English while Russian astronauts spoke only Russian. Each astronaut then only had to be able to understand, but not to produce, the other tongue. This made communication much easier and more accurate, since receptive language ability can reach high levels of efficiency in a much shorter time.

I have noticed this same pattern in my own language experience. I can interpret English, Russian, German, Spanish, French, and Italian enough to be able to understand movies in these languages without subtitles or to follow conversations with reasonable ability. On the other hand, I believe that I speak English fluently, Spanish moderately well, German at a lower level, French with minimally understandable competence; my Russian and Italian productive language make me sound like a two- or three-year-old as far as linguistic fluency is concerned. Thus, like all human infants, my receptive language is many times better than my productive language.

Dogs certainly have the sound discrimination ability to develop receptive language. They can actually pick up quite subtle nuances in human word pronunciations. One example of this involves the ethologist Victor Sarris. He had developed a fondness for the sound of his own name, and for this reason he gave his three dogs names that rhymed with Sarris, calling them: Paris, Harris, and Ariss. While one might have expected that this would cause confusion, it did not. Each dog responded to its own name accurately, and none seemed to bear a grudge toward their master over his odd naming practices.

One must not underestimate the receptive language ability of dogs. The fact that they are not able to produce human sounds to communicate with us doesn't mean that they do not comprehend human words. A dog can prove that it comprehends when it responds to spoken words appropriately. The dog can obey a spoken command, or produce an intelligent or appropriate behavior in response to our spoken message. Everyone who has ever lived with a dog knows that they quickly learn to respond to a number of human words. As an example, let me give you a mini-dictionary of my own three dogs' working vocabularies. This should provide some sense of the scope of a typical dog's receptive language ability, though it is certainly not the upper limit of what they are capable of learning.

Some of the words and phrasings my dogs learn are quite idiosyncratic to me, reflecting my lifestyle and the way that I interact with my dogs. Some of the words may also not be responded to by all three of my dogs, since that depends upon their age and present level of training. It is also true that my partial list includes only words that I deliberately use to get a response rather than other words that the dogs may understand but aren't formally required to respond to. Each word in this sampling is presented along with the actions which demonstrate the dog's comprehension.

Away: The dog responds to this command by moving back from whatever it was investigating or attending to.

Back: I use this only in the car. In response, the dog moves from the front to the backseat area.

Bad dog: This is a term of displeasure. The dogs recognize the implicit anger and usually respond by submissively cringing and sometimes by leaving the room.

Be close: I use this phrase when walking my dogs. In reply, a dog that has been lagging too far back will move up to a position closer to me.

Be quick: This is taught during housebreaking. Upon hearing it, the dog will start searching for a place to eliminate, even if this is only a token leg lift to please me.

By me: This is a multipurpose command used in daily activities to get a free-ranging dog to return to a position close by my left side.

Collar off: This useful expression causes the dog to respond by lowering its head to allow its collar to be slipped over and off more easily.

Collar on: Obviously, this is the companion phrase. In answer, the dog lifts its head, pointing its muzzle up, to allow the collar to be slipped on with less effort.

Come: The basic recall command.

Den: One of many "go-to-someplace" commands. In this case, it directs the dogs to go to my office at home to wait for me.

Do you want to play? This phrase causes the dog to circle, bark, and emit play bows in preparation for some fun and exercise.

Down: This causes the dog to lie down immediately, without changing locations.

Downstairs: When it hears this, the dog responds by going down a set of stairs in front of it.

Drop it: This defensive expression is taught to my dogs as puppies, when they are most apt to pick things up in their mouths that might be harmful to them. In response, the dog spits out whatever it has in its mouth onto the ground.

Excuse me: A useful phrase that I use when one of my dogs is blocking my path of movement. In reply, the dog gets up and stands aside to allow me to pass.

Find glove: A command that is part of formal dog obedience competition training. In response, the dog searches back across the ring to retrieve a glove that has been dropped earlier.

Find it: Another competition command. It tells the dog to find the one item with my scent which has been placed among a group of other items carrying the ring steward's scent.

Front: The obedience version of the more casual recall command, "Come." When I tell the dog to "Come," simply arriving near me is all that is required. When I tell the dog "Front," however, he is supposed to return and then sit squarely in front of me until I tell him what to do next.

Give: This word is used when I want to take something out of my dog's mouth. In response, the dog releases the pressure on any item it is holding in its mouth, so that I can remove it easily.

Give me a kiss: In answer to this, the dog licks my face.

Give me a paw: When it hears this, the dog will lift the paw nearest my hand to have nails clipped or paws toweled dry.

Go back: This command is always given with a hand signal to indicate direction. When it is given, the dog moves away from me in a straight line along the direction indicated until I tell him to stop.

Good dog: The general purpose term of praise that usually causes tail-wagging in pleasure. It is interchangeable with "Good boy," for my all-male dog collection.

Heel: The traditional command to get a dog to walk under control by my left side.

Hugs: This is a silly command, but I like it. I use it to get my dogs to jump up in front of me, with their legs on my thighs, to allow me to pet them without bending.

In: The dog responds to this command by going through an open door or gate in the direction indicated by my hand motion.

Jump: I use this word to get a dog to leap over some object or obstruction I'm pointing to.

Kennel: In response to this command, the dog goes into its kennel.

Lead on: Another of those utilitarian phrases that make life with the dog easier. In answer to "Lead on," the dog lifts its head to provide access to collar ring. "Lead off" produces the same response.

Let's go: Really the casual form of the "Heel" command, where the dog's only requirement is to stay reasonably close to me as I walk. The dog can be a short distance in front or back of me and doesn't have to sit when I stop.

Loose: A play command that tells the dog it is free to chase an object I have thrown.

No: This command is always given in a loud, sharp tone of voice. The intent is to have the dog freeze and stop all ongoing activities. To get that freezing response, the first few times that I use this command with a new puppy, I accompany it with a loud, sharp sound. Banging a pot against a counter works well; slapping a wall or a table, stamping one's foot on a wooden floor, even throwing a book down flat on the floor will also work. This command is extremely useful to keep your dog out of trouble. A shout of "No" can freeze a dog who is approaching a frightened child or a dangerous situation. Once he has stopped his approach, the command, "By me," brings him to my side, where I can control him with my hands and supervise any subsequent events.

Open your mouth: I use this when I'm cleaning my dog's teeth.

Out: A limited-use command, which simply tells the dog to leave its kennel or whatever other enclosure it is in, such as my car.

Pick up your toys: A handy housekeeping phrase that causes the dog to search the room for dog toys and bring them to me.

Playtime: This word is a release, which indicates that an exercise is finished or that the requirements imposed by the last command have now terminated. The word causes the dog to break from position and come for praise; in other circumstances, the dog may dance about a bit or explore the room or greet nearby people or other dogs.

In the past, with other dogs, I used the word "Okay" as my release, since it seemed so natural. Unfortunately, "Okay" is a common phrase in the language and people may often say it with great enthusiasm, so that my dogs thought that other people were giving them their release. Once, at a dog show, someone yelled: "Okay!" in joy when their dog won best of breed in the conformation competition. Unfortunately, I was in the middle of an obedience trial in the next ring. My cairn terrier, Flint, was supposedly in a down and stay position for five minutes, with me out of sight with the other competitors. The sound of his release command so close was just what he was waiting for, so he popped up happily and began trying to introduce himself to the other dogs, who were still dutifully doing their exercise. Starting the very next day, I dropped "Okay" as my release word and substituted "Playtime." At least this word is used with much lower frequency in general conversations and virtually never in the dog obedience ring.

Quiet: This stops a dog's barking—at least for the moment.

Relax: This slows my dog down to relieve tugging on the leash when we're walking.

Roll over: Well, everybody has to teach the dog some "tricks" to keep their children or grandchildren happy. In response, the dog rolls on its back for a belly rub.

Seek: This is part of formal tracking training, which accompanies the presentation of a scented object taken from somebody. In response, the dog tries to follow the scent trail to find the individual.

Settle: A command usually accompanied by a hand signal indicating place. It means that the dog is to settle down and remain quiet in that given area. The dog may sit or lie down, or even stand or move occasionally, but it may not engage in any active actions.

Sit high: Simply another command to cause the dog to perform a silly little, common "trick." In response to "Sit high," the dog sits on its hind legs, with its front legs off the ground in the traditional begging position.

Snuggle: This is a word that I teach all my dogs when they are puppies. At the command "Snuggle," the puppy (who is being carried) drapes its head over my shoulder and rests there.

Speak: Again, a simple little traditional dog trick command. It causes the dog to give a single bark in reply.

Stand: This word has a different effect depending upon whether the dog is walking, sitting, or lying down. If the dog is moving, the word "Stand" causes it to stop moving and stand facing in the direction it was walking. If the dog was sitting or lying down, it now rises up, takes a step or two forward, and then stands quietly in the direction it was facing.

Stay: This very specific command tells the dog that it is supposed to remain in its current place and position until released.

Steady: I use this as a variant or reinforcement of "Stay." It is only employed during grooming, when I'm using the dematting brush or otherwise pulling at the dog's hair, which may be uncomfortable. The word "Steady" causes the dog to lock and hold its position despite any momentary discomfort or pulling.

Swing: My command to get the dog to circle around behind me and then sit near my left leg.

Take it: The formal command for my dog to retrieve an object.

Time to clean your eyes: In response, the dog places its head in my left hand so that I can perform the ritual of cleaning the tear stains from around its eyes.

Towel time: On hearing this, my dogs go to the center of the room (usually the kitchen) to wait to be dried off after a walk in the rain.

Up: This word is usually accompanied by a hand signal. The hand signal (a pat on, or pointing to, an object) indicates the surface the dog is supposed to jump up onto.

Wait: A much looser version of the command "Stay." It is meant to cause the dog to temporarily stop any current activity and remain roughly in position while continuing to watch me for further instructions.

Watch me: This is an attention-getting command. It alerts the dog to keep its eyes on me, since I will be giving a command in a moment or two.

Where's your ball? One of several "Find the object" phrases that my dogs respond to. If the object is accessible and small enough to carry, the dog will usually bring it back to me; otherwise, he will usually stand near it and bark.

Who wants a cookie? All of the dogs within earshot of this question immediately run to the kitchen counter to wait for a dog biscuit.

Who wants a ride? When I'm outside the house, this question causes the dogs to run to the van and wait to get in.

Who wants some food? Interchangeable with "Chow call," this causes the dogs to run to the kitchen and face the place where their food bowls are put out, in anticipation of a meal.

Who wants to go for a walk? The dogs go to the front door and wait.

Now, this list of words and phrases is most definitely incomplete. I've only given my most commonly used vocabulary items and have not included words that produce untrained responses. Thus, the word "bath" spoken by me in conversation with my wife can elicit different responses depending upon the dog doing the listening. My former cairn terrier, Flint, would respond by looking for someplace to hide. My Cavalier King Charles spaniel goes to the door of the bathroom to wait for the inevitable, while my flat-coated retriever simply becomes alert, watching the scene to see if this word has any implications for his future.

Over time, I have become aware that the dogs respond to many other words that I use but have never deliberately taught them. I have come to notice that at the phrase "dog class," they begin to hover excitedly near the door and watch the closet where I keep my training equipment. Recently, I noticed that the word "office" has developed a meaning for them, but the action that it causes depends upon where we are at the time, and is different for different dogs. Thus, if we are out at our farm, and I tell my wife, Joan, "I'm going into my office to do some work," the dogs begin to migrate in the direction of my office, where they will lie near me while I write. When I'm in the city, however, this same phrase may mean that I'm going to sit down in my home office and write, or it may indicate that I'm about to leave the house to go to my university office. My spaniel responds the same way he does in the farm, and afterwards he can be found resting in the footwell of my desk. My retriever, Odin, finds my briefcase and stations himself near it. He has figured out that I will usually pick up my briefcase before I move either toward my office in the house or toward the front door to leave.

Other habitual phrases also seem to be interpreted, such as, "I think that I'll go to bed now," which causes Odin to climb the stairs to our bedroom and settle down on his pillow beside the bed. I'm sure there are additional common language messages that the dogs regularly respond to, but I have not yet isolated the behaviors which indicate they are correctly interpreting the words.

I have had reports about some bright dogs that have learned receptive language so well they've become pests. Rita, the owner of a white standard poodle named Tony, found that she couldn't use certain words in casual conversations in front of the dog without his responding to them. The mere mention of the word "walk" in a sentence would cause Tony to run to the front door, barking at the prospect of going out. Similarly, mention of "ball" would result in his frantically searching for his toy, while "eat" would incite him to hover expectantly in front of the refrigerator. There seemed to be about half a dozen words that he responded to in this way. Since it got to be quite annoying to Rita and her husband, they began to resort to spelling out the words. Thus, she might say, "Do you want to take the dog for a W-A-L-K?" But Tony's receptive language was so good that he quickly learned the sounds of the spelled-out words, and would respond to them in the same way that he responded to the spoken words themselves.

How many words or phrases can a dog learn? There is a lot of debate about this. If we restrict ourselves only to words and sounds, some psychologists, such as J. Paul Scott, suggest that it would not be unusual for an average dog to distinguish nearly 200 spoken human words, which would put the dog's language ability at about that of a human two-year-old child. Some dog trainers claim that dogs can learn an even greater number than this, perhaps in excess of 300 words. I received a letter from a German dog trainer who claimed that he had taught a German shepherd to respond to close to 350 words. "These words do not have to be one- or two-word commands," he wrote, "but can be part of a sentence. The dog just picks out the important part and does what he is required to do." There is certainly evidence that this is the case.

One historical example of dogs seeming to filter human language for bits and pieces that make sense to them comes from a time when many dogs were specifically bred as a compact source of cheap power. Everyone knows that dogs have been used instead of horses, mules, or other animals to carry materials in backpacks and to pull loads in sleds or small carts. The fact that dogs have been used as a power source to drive light machinery is not as well known. For centuries, dogs had a special place in the kitchen of large households. At that time, meat was usually cooked over an open fire on horizontal spits. These spits needed to be rotated continuously to cook the meat evenly. The tedious job of turning the meat was given to a special breed of heavy, long-bodied and short-legged dogs, appropriately called "Turnspits." They were placed in an enclosed wheel, which looked like a larger version of the suspended wheels you sometimes see in hamster or rat cages. As the dog walked in this wheel, it turned with each step. This generated the rotary motion needed to rotate the metal spit that was attached to its central hub. A house might have several Turnspits, and each dog might be required to work the wheel for a number of hours. Turnspit dogs were also used generate the motion needed to churn butter, grind grains, pump water; there is even a patent from that era for a dog-powered sewing machine.

These dogs were not always confined to the kitchen but often had other more pleasant tasks. When one or more of these dogs was not needed as a source of power, Turnspit dogs were often taken to church, where their job was to serve as footwarmers. One Sunday, the bishop of Gloucester was giving a service in Bath Abbey. He drew his text from the tenth chapter of the Book of Ezekiel. He was a fiery speaker who put great emphasis in his words. At one point he turned to the congregation and shouted, "It was then that Ezekiel saw the wheel." Up to this point, the dogs in the church had been resting quietly at the feet of their masters. They were clearly monitoring the flow of words, however, perhaps seeking to extract any information that might pertain to them. At the mention of the word "wheel"—the Turnspit's dreaded workplace—there was a sudden reaction. One witness reported that a number of dogs "clapt their tails between their legs and ran out of the church."

The important place that dogs have in our lives is illustrated by the fact that we give them names. This endows each dog with an individual identity and indicates their significance. It also assigns a particular sound to each dog. What is most interesting is that dogs come to respond to the sound of their names. In the wild, social animals seem to have no need for naming. Each animal knows its place in the group and interacts without any vocal identifying label. It is only in the world of people that dogs need to learn this bit of receptive language.

For human beings, names are very special things. According to the Bible, one of the first tasks that God gave to Adam was to assign a name to each living thing. In many cultures, the name of a person contains the essence of that person, and speaking that name may even give someone control over that person in a magical way. There are several cultures, for instance, in which each child is given a "true name" at birth, but this name is never spoken, and is known only to the child and those who gave it to him, to prevent any form of magical control. Instead, children are also given another name by which they will be called in their everyday life.

When it comes to animals, only very special creatures are given names. A farmer doesn't normally name his chickens or his beef cattle. To give a name to something or someone is to acknowledge that thing as an individual, with an identity and personal feelings. People who don't acknowledge dogs as individuals will tend to use a depersonalized label when referring to it. A person who says, "The dog is hungry. Feed it," shows the same lack of caring that would be shown if they were speaking about their son or daughter and announced, "The child is hungry. Feed it." People who care use the individual's name. It is Lassie or Sarah or George who is hungry if we care about them.

The Eskimos go one step further when thinking about their dogs. It is their belief that dogs don't have a soul unless they are given a name. The special names that give a dog a soul are actually human names, and often the name of a deceased relative. Only a few dogs are given these precious names. These lucky dogs are taken into the home and are usually better fed and treated more like pets than the dogs who simply pull sleds and work. The Eskimos do need some way of identifying each of the common working dogs, so each of these dogs receives a sort of label. These may be based upon their physical characteristics, such as Gray, Blackie, Long Tooth, Spotted Tail, Brute, or upon their abilities or behavior, such as Runner, Sleepy, Happy, or Braveheart. These identifying tags, however, aren't really names, and according to Eskimo tradition, do not carry with them a soul.

Even if we disregard any mystical or religious beliefs about the matter, dogs' names are a vital aspect of their lives. Remember, a dog lives in a sea of human language sounds. However, its vocabulary is relatively limited, much like that of a young child. The dog's first task, then, in interpreting our language is to try to figure out which of the words that it does understand are actually being directed to it personally, and which are not. Thus, a casual comment that you might make to a family member, such as, "Why don't you come over and sit down and watch television with me," could be a real problem for the dog. For my dogs, that simple sentence contains three words that each of them knows well—and these are the common obedience commands, "Come," "Sit," and "Down"—as well as the attention-getting word "Watch." Suppose that a dog was sitting in the room with you when you spoke that sentence to your family member. Would you really expect any bright dog to suddenly begin to execute that sequence of behaviors by first coming to you, then sitting, then going into a down position and looking directly at you? If not, why not? How does the dog know which of the many words that you say are directed at him, and should be immediately responded to, as opposed to those which do not affect him at all?

One of the ways that dogs can interpret whether human word sounds are directed at them is through our body language. Obviously, if I'm looking directly into the dog's eyes and have its full attention, then there is no ambiguity about the words "Come," "Sit," or "Down." Under such conditions, the dog knows that these are commands that are clearly directed at him and he should know that you are expecting him to respond to those words. In the absence of that sort of explicit body language, however, the dog's name becomes the key to his understanding. In effect, a dog's name becomes a signal which tells him that the next sounds that come out of his master's mouth are supposed to have some impact on his life. It is as if the dog's name translates into something like "Listen carefully, this next message is for you."

Because of this, we have to be very precise when talking to our dog. Each time we want it to do something, we should start off with its name. That means that "Rover sit," "Rover come," or "Rover down," are all examples of proper dog talk. On the other hand, "Sit, Rover," or "Come here, Rover," are examples of bad grammar when speaking to our dog. The reason that they are bad is simply because the words that you want the dog to respond to will have disappeared into the wind well before it has been alerted that the noises you are making with your mouth are addressed to it. When you say, "Sit, Rover," since nothing meaningful follows his name, you may well end up with a dog simply staring up at you with that "Okay-now-that-you-have-my-attention-what-do-you-want-me-to-do?" look that we all have seen so many times. The reason the dog is looking at you is that you have now caught his attention by using his name, and he is waiting for information about what you desire of him. After a few seconds of this exchange of gazes, you might well repeat the command in an annoyed tone. "I said sit, you stupid dog." The dog now sits, but it was the human that was being stupid.

Pedigreed dogs usually have two or more names. Their first name is the one that is registered with their kennel club. Registered names are usually marvelously pompous and/or meaningless labels, such as "Remasia Vindebon of Torwood," "Flatcastle Shadow on Wild Water," "Blacklace Moonlight Romance," "Parkburn's Rainy Day Woman," or "Solar Optics from Creekwood." However, the dog's most important name is its "call name." After all, you really don't want to be standing out in your backyard yelling, "Tollbreton Coranado Dancer, come!" Most people will find a short name to give a dog for everyday interactions. The dog's call name becomes its own unique, solely owned name, which is the one we actually use when we talk to them. For my dogs, they are names like Wiz, Flint, or Odin. Over the years I have found that two-syllable names seem to roll off my tongue more easily and tend to produce a better response. Thus, Wiz is actually called Wizzer most of the time, while Flint obtained the fake Latin name Flintus. I like there to be some link, no matter how tenuous, between the registered name and the call name. Thus, "Tollbreton Coranado Dancer" was given the call name "Dancer," while "Koy's Abracadabra Alchemist" got the call name "Magic."

Some people use their dog's call name to try to create a specific impression. In the world of professional athletics, people are often trying to establish an image of themselves as being tough and dominant. These people often select big, tough-looking dogs, such as Rottweilers, bull mastiffs, Doberman Pinschers, and Great Danes, to emphasize the fact that they themselves are powerful and tough. Such dogs are usually adorned with the proper accoutrements to reinforce that image, such as heavy leather collars with studs. In addition, the dog must have the right name. Herschel Walker, who became the all-time leading yardage gainer in professional football in 1995, has a Rottweiler named Al Capone. Some other names of dogs owned by professional athletes include: Slugger, Rocky, Hawk, Ghost, Jagger, Trooper, Rocket, and Shaka Zulu. A dog with the name of Fluffy, Honey, or Fifi just won't work.

Does giving your dog a dominant-sounding name cause people to view you yourself as more tough and dominant? Well, this is not really clear. However, giving your dog a hard or threatening name certainly seems to affect the way in which other people will view and react to your dog. I verified this in a laboratory experiment. People were given the description of a dog which said something like:

We are interested in your ability to determine the personality and intentions of dogs by simply looking at their behavior. We will show you a brief video clip of a dog named Ripper, interacting with a person. Watch the dog carefully because we will be asking you some questions about Ripper's behavior.

Ripper could be changed to some other tough name, such as Killer, Assassin, Butcher, Gangster, and so forth, or to more positive names such as Champion, Braveheart, Happy, Lucky, and so on. The video consisted of clips taken from a television series which starred a German shepherd dog. The short sequence consisted of a man walking; then, from off screen, the dog runs up to him. There is a close-up of the dog barking at the man, followed by the dog jumping up and placing its paws on the man's shoulders. The man pushes the dog away, and the dog runs barking out of the scene.

A total of 291 people viewed the video after receiving the brief description which contained what was supposedly the dog's name. Afterward, they were given a list of words and told to check off those they felt best described the dog they had just seen. These words included typical adjectives used to describe positive attributes such as friendly, sociable, cordial, or playful, and also negative attributes such as aggressive, threatening, hostile, or dangerous. When the dog's name was tough, as in Assassin or Butcher, they were much more likely to describe the dog's behavior as being hostile or menacing than they were when the dog had a more positive, less threatening name.

One of the most interesting aspects of this study is that when these people were asked to describe the events they had just seen, those who had heard the dog described with a hard name such as Slasher were more apt to say things like: "The dog saw a man and didn't like him. The dog barked at him and tried to jump up on him to make him go away, but the man pushed him off before he could be bitten and the dog ran away." People who had had the dog described as having a more positive name, such as Happy, were more apt to describe the very same scene as: "The dog saw a man coming and ran out to say hello. The dog barked and jumped up to try to get the man to play. Then the dog ran ahead of him to lead him home." Remember, both of these people had seen exactly the same video clip. The only difference was the name of the dog.

This kind of data makes it clear that the name you give to your dog makes a statement about that dog to other people who are going to come into contact with it. Obviously, the choice will be more important for a larger dog. Somehow, I doubt that a Pekingese, Chihuahua, or a Maltese will produce a sense of dread or threat, even if it is named Exterminator, Killer, or Beast.

Virtually any word can become a name. Simple words, gleaned from the dictionary, like Promise, Spice, Skylark, or Runner, all seem to turn into pleasant call names for dogs. Atlases have some interesting possibilities, such as Oxford, Newgate, or Congo. If people have special interests, lists of technical terms can produce some workable names. I know a lawyer who has a dog named Hearsay, a geologist with a dog named Granite, and a recreational sailor with a dog named Rudder. The names can be quite arbitrary. For example, the singer Frank Sinatra once gave the actress Marilyn Monroe a white poodle named Maf. He told her "Maf" was short for "Mafia," which we all know is an organization with which he had no association.

The top ten dog names in the United States and Britain have been remarkably constant over the past decade or so. Starting with the most popular, these names are:

	MALES	FEMALES
1	Max	Princess
2	Rocky	Lady
3	Lucky	Sandy
4	Duke	Sheba
5	King	Ginger
6	Rusty	Brandy
7	Prince	Samantha
8	Buddy	Daisy
9	Buster	Missy
10	Blackie	Misty

For me, the great surprise was that the name Snoopy (the dog made famous by Charles Schulz's popular comic strip "Peanuts") was not very high up in the list. On the other hand, I did find it in the top ten names for cats!

In addition to their registered and their call name, all of my dogs have a third name. This is a group name, which for me is "Puppy." For each dog, this is their alternate name; thus, when I yell, "Puppies come," I expect all of my dogs within earshot to appear at a run. A friend who has only male dogs uses the word "Gentlemen," while another (a former officer in the Army Tank Corps) uses the group name "Troops." From the dog's point of view, this group name is just another sound which means, "Pay attention, the next sound is for you to obey."

The kennel club allows only one change of a dog's registered name during its lifetime. Dogs, however, are more flexible, as long as there is a chance for them to learn their new label. Thus, dogs adopted from pounds or shelters often have no known name, but quickly learn whatever new name is given to them.

Sometimes, even when living consistently with one family, a dog will find its name changing. This usually happens when a casual nickname is applied to the dog and somehow manages to stick around for a long while. For example, my daughter by marriage, Karen, had a dog who was originally named Countess. Her call name became Tessa, but she also learned to respond to Tess, Tessa Bear, T-Bear, and Bear. Tess handled all of these changes of title with the same aplomb that we might have if a loved one who had always called us "Darling" decided to start calling us "Honey" and sometimes "Dear."

Tessa's name-change problems were minor compared to those some other dogs have experienced. For example, there is the case of the Skye terrier owned by Robert Louis Stevenson. Stevenson is best known for writing such classics as *Treasure Island* and *The Strange Case of Dr. Jekyll and Mr. Hyde*. His little dog was initially named Woggs, which was then changed to Walter, then modified into Watty, then transformed to become Woggy, and finally ended up as Bogue.

Any sound that is consistently used with a dog can come to be its name, at least for a while. I had an interesting experience with one dog, a Siberian husky named Polar. I had been invited as a special guest speaker at a scientific conference being held at a ski resort. I was to be housed in a cabin shared with Paul, one of the conference program directors. Paul lived within driving distance of the resort, and had brought Polar with him. He knew that I liked having dogs around me at all times, and thought it might help me get over the separation pangs I have when I'm on the road and away from my own puppies.

It was interesting watching Polar and Paul interact. Although Paul clearly loved the dog, he was having some trouble controlling this rambunctious, bouncing ball of fur. As soon as the car door opened, Polar dashed out. Paul yelled, "No!" and the dog obediently came back to his side. When I went to greet the dog, he jumped up on me, and Paul again brought him back down to the floor with a sharp "No!" That evening, as Paul and I sat chatting over a drink, Polar began nuzzling him to try to get one of the pretzels we had in the bowl between us. Again a quick

"No," and Polar settled down with a sigh. Later that night, there was some commotion on Paul's side of the room. Polar had tried to snuggle his way onto Paul's bed and was pushed off with another sharp "No!" In the morning, the first sounds I heard were from Paul telling Polar, "No, it's too early. I don't want to get up yet." Then, a few minutes later, "No, let me sleep. I'll let you out in a while."

Later, over dinner, Paul confided that he sometimes felt he didn't really have the dog under control for much of the time. "For example, there are times when I don't even think that Polar knows his own name."

"Polar knows his name," I told him; "however, you might not." In response to his puzzled look, I continued, "We'll run a little experiment when we get back to the cabin tonight."

Later that evening, back in the cabin, I instructed Paul to stand in the kitchen area, and I took Polar with me out on the deck just off the bedroom. I was petting Polar, who seemed contented at the attention that he was receiving, when Paul, standing in the kitchen, shouted (as prearranged): "No!" Polar stood up and quite obediently trotted off to his master. On the basis of what he had experienced during his life, the sound that he had heard most frequently associated with consequences for him personally was "No." In Polar's mind, then, "No" was his name!

We have been talking about the receptive language ability of dogs, specifically, their ability to understand aspects of human language. The examples I have used come from fairly typical dogs living in an average family situation. However, there are many reports and stories about dogs with truly amazing abilities to understand what people say. If I believe these accounts, then my dogs are really barely out of first grade in comparison to the college-level performance of these other talented canines. It may be worthwhile to take a look at the receptive language of some of these highly educated dogs, since there are some surprising secrets to be found in their achievements.

4

Is the Dog Really Listening?

I suppose that it was sometime in the 1960s that I had the opportunity to see a demonstration by Charles Eisenmann and his dogs. "Chuck" Eisenmann, as he was known, had started out as a professional baseball player, and then, through a series of career twists, had ended up as a dog trainer and handler. Although his dogs appeared in many Hollywood films, he is best known for the long-running TV series *The Littlest Hobo*, which starred his German shepherd dog, London. London had several sons, including Little London, Toro, and Thorn, who often served as London's doubles for certain stunts. All were touted to be remarkably accomplished dogs. Eisenmann claimed that these dogs knew several hundred individual words, and had the linguistic comprehension of an eight-year-old child.

The demonstration that I attended had been arranged by, or was sponsored by, a local television station, and there were some cameras around to tape or film some of the proceedings for later broadcast. Eisenmann introduced his four dogs and explained that most dogs were simply trained to form habits, such as lying down when they heard the word "Down" or coming when they heard the word "Come." In essence, they were simply associating a sound with an action. He went on to explain that he trained his dogs using what he called "the Intellectual Method." The crux of this method was that it supposedly forced his dogs to think and to learn the basic elements of spoken language. His teaching technique appeared to be similar to the way in which we teach language to children. Instead of associating a single word with an action or a concept, he varied the wordings that were used, and presented the same problem in a variety of different settings. He demonstrated the end product of this training by showing that London not only responded to "Down" but to alternate phrasings such as "Kindly recline on the floor" or "Assume a prone position."

These dogs certainly did have good language comprehension. When

Eisenmann spoke to them in a natural conversational tone, using casual everyday wordings, the dogs showed that they understood by doing whatever he told them, whether it involved opening or closing a door or turning a light switch on or off, or a variety of similar actions. I remember being most impressed by the dogs' ability to go over to a group of objects and select any particular item their master named. My credulity became a little strained, however, when Eisenmann claimed that his dogs would respond just as well when he spoke to them in French or German. I suppose that, because it was California, it was inevitable that someone called out, "How about Spanish?"

Eisenmann replied, "I don't know if the dogs understand Spanish. Why don't we try? Give me a simple command in Spanish."

He was told that *Cerra la puerta* meant "Close the door" in Spanish, and with that knowledge, he turned to the dog and said, "London, *cerra la puerta!*" The dog stood, and although he appeared to be a bit slower and more hesitant, he moved to the door that had been used in an earlier demonstration of his ability to open doors on command. London looked back at his master, then pushed one paw forward, and the partially open door swung closed. The audience broke into a loud, raucous cheer of approval.

This demonstration truly bothered me. I believe that dogs can be trained to understand a good deal more language than most people give them credit for. I also believe that they can learn commands or words in several languages, much the way that humans do. Yet I know that each word in each new language requires some form of learning before that comprehension can be demonstrated. For example, my knowing that the word "dog" in English, *"Hund"* in German, and *chien* in French, all mean the same thing, certainly doesn't equip me to know that the word *perro*, in Spanish, also means dog. Until I am given some information about the translation, I can't be expected to understand a word's meaning. How then could the dog London understand a command the first time he heard it, if, as it seemed, his master didn't know the language and had never taught it to him? The habit of questioning is part of being a scientist, and this leads to a hesitation in accepting all things at face value, even if it is something that I *want* to believe. At that moment, caution flags began to wave in my mind and I viewed the performance with a bit more skepticism.

My worries soon increased. Eisenmann was explaining to the host that many people believed that dogs were color-blind. He could prove this was not the case.

"London, point to something in this room that is red," he said.

The dog got up, moved across the room, and pointed his muzzle at a red coffee mug next to the host's hand. When asked to point to something blue, he pointed to a blue chair; finally, when asked to point to something yellow, London moved toward the wall and pointed his nose at a yellow curtain.

While the audience cheered, I began to fret. Once again, London's performance was too good to be true. Dogs' eyes are different from human eyes and it is in the area of color vision that dogs show the greatest relative deficit. Dogs are not actually color-blind, if we mean that they see the world only in terms of shades of gray. Scientists have been able to show, using special training techniques, that dogs probably see the world in shades of gray, green, and brownish-red. The fact that dogs can be trained to discriminate these colors shows that they have some color vision ability. The flip side of that coin is that these training procedures are very laborious and exceptionally difficult, which means that color vision is probably of little importance to the dog. From a biological viewpoint, color vision is only important for animals that function primarily during the daylight hours and have a very varied diet. In that case, color vision is an aid to finding and identifying various things that could be food. Dogs, being naturally twilight and dawn hunters, have much less use for this capacity and do not seem to use it spontaneously.

Let us suppose for the moment that Eisenmann had managed to make colors important for London, so that he would use his limited color vision capacity whenever asked to. Even then, since dogs discriminate the world in greens and reddish-browns, the fact that London could also identify a blue and a yellow, which are colors that his eye cannot distinguish between, is amazing—and unlikely.

Now I began to watch very carefully. Eisenmann was continuing with his "performance." Next, he told London to "find something which has printed words on it." The dog looked carefully at his master as he spoke the words and then walked over to a poster on a stand and indicated it. Later, the dog pulled a pencil off a low coffee table when asked to "bring me something that I might write on paper with." He amazed the audience with his ability to even spell. When asked to "bring me a pair of g-l-a-s-s-e-s," London walked over to the host; while the crowd giggled, he gently removed his glasses and carried them in his mouth to his master.

The language comprehension shown by this dog simply was beyond the realm of reason. If Eisenmann's contention that all dogs could be taught the comprehension of an eight-year-old human child was true, then why do we have so many "underachieving" dogs in our world?

In the end, the secret of London's performance was revealed to me by Professor Carl John Warden, one of the most respected comparative and evolutionary psychologists of the early twentieth century.[1] While Warden was at Columbia University in New York, he had the opportunity to test a German shepherd dog named Fellow, who was owned by Jacob Herbert of Detroit, Michigan. Herbert was a dog breeder who had selected Fellow out of the many dogs he had bred as being one of the most intelligent. As a sort of personal project, he set out to try to teach Fellow as much human language as he could. He had hit upon the same procedure that Eisenmann had, which simply involved talking to the dog continuously, in the same instructional way that we might talk to a young child we were living with. Herbert believed that his dog knew somewhere in the vicinity of 400 words and understood them in much the same manner as a child would under similar circumstances. Herbert wasn't crediting Fellow with full language ability, but certainly felt that he had formed connections between certain words and certain objects or actions.

Herbert recognized that he was not an expert in animal behavior, and he honestly wanted to find out just what Fellow's language capacities were. He contacted Professor Warden, and they arranged a test of Fellow's abilities for his next visit to New York City. This first set of tests was conducted in Herbert's hotel room. Warden and an associate, L. H. Warner, approached the tests with what Warden called a "chronic skeptical attitude." They soon became quite impressed by the fact that the dog performed remarkably well, responding to a wide variety of commands. Warden was surprised (in much the same way that I was surprised at London's performance) by the fact that the dog's owner didn't try to use identical phrases when he was asking Fellow to do things. Furthermore, all of the commands were given in a casual speaking voice, as though Herbert were simply having a conversation with the dog.

The psychologists tried to determine if there was something other than understanding the words that could account for Fellow's excellent performance. First they tried having Fellow respond to a list of spoken commands in a completely different order than his master had thus far used. He was just as accurate in that test, which meant that there was no set "routine" that was being followed. It seemed that Fellow was responding to the specific words, since when his master deliberately varied the tones that he spoke to the dog, from a high or a low pitch to a monotone, it made no difference to the dog's performance. To try to make sure that some sort of secret signaling wasn't involved, Herbert was stationed in the bathroom of the hotel suite, with the door closed, so that he was out of sight. Although Fellow's performance under these conditions was not perfect, he was correct most of the time. This was surprising, since this was a new situation for him, and word sounds were noticeably a bit muffled by passing through the door.

Herbert and Fellow were then convinced to come to the Columbia University campus, where they were more formally tested. In these tests, both Herbert and the psychologists were concealed behind screens, observing the dog's behavior through small slits. In a set of retrieving tests, it became clear that Fellow knew the meaning of many common words. Common items that Fellow knew included key, brush, glove, package, pillow, water, milk, shoes, hat, coat, stick, ball, post, money (paper), dollar (silver coin), lady, gentleman, boy, girl, puppy, and so forth. Other tests suggested that he knew the names of various body parts, such as foot, head, mouth, paw, and lap (of a person). Remarkably, he could also distinguish the size of objects, as in the difference between a "big boy" and a "little boy."

During the testing session, the psychologists verified that Fellow could reliably respond to fifty-three different commands, sentences, and phrases, even though his master was not visible. These ranged from simple action commands like "Sit," "Roll over," "Turn your head," "The other way" (causing the dog to turn his head in the other direction) to the more complex "Go and take a walk around the room," "Go outside and wait for me" (causing the dog to leave the room and wait outside near the door), "I don't trust him" (which caused the dog to bark and threaten an attack), and many more. In addition, there were some fairly complex ideas that could be obeyed accurately, such as, "Go along with the gentleman (or lady)," while "Stop," "Quit that," "Never mind," or "Still" all resulted in the dog stopping whatever activity he was engaged in. "Do that once more" resulted in the dog repeating whatever set of actions he had just done before.

Although Fellow's performance was extremely impressive, there were some limits to his ability. There were certain commands he could carry out with perfect accuracy when his master was present, but did not seem to understand when Herbert was hidden by the screen. When these commands were analyzed, it became clear that they all had something in common. These commands had two components. First, there was an object to be identified, and then there was a location that had to be oriented toward. Thus, Fellow would respond accurately to "Go and find Professor Warden," when his master was present, but not when he was not visible. Other phrases which produced similar difficulties when the dog's master was out of sight included "Go and look out of the window," "Now go to the other window," or "Go and jump up on the chair [or table or other appropriate object]."

Since each of these commands required the dog to first orient himself in a particular direction and then perform some action with the object there, Warden guessed that there might be some subtle visual signal being given by Herbert to the dog. Since Herbert seemed quite honest and was interested in objectively determining how much language Fellow knew, it seemed likely that this signal would have to be a fairly natural and unconscious movement. The most obvious movement of this sort is simply a head turn. If a person asks for something, as in "Please bring me the telephone," it is quite natural to glance in the direction of the table on which the phone rests. This kind of head turn, or glance of the eyes, happens almost automatically whenever we talk about an object that is visible from where we are at the moment.

To test this, Professor Warden had Herbert deliberately try to mislead Fellow. Thus, Herbert was instructed to say, "Go over to the door," while looking at a table across the room. Fellow responded immediately when he spoke; however, the response followed the direction that Herbert was looking, rather than the words, since the dog got up and moved directly to the table. When Herbert looked at the window but asked, "Jump over the chair, good dog," Fellow went straight to the window toward which his master had turned his head. Next, the dog was asked to "Go put your head on the chair." Instead, Fellow jumped up on the table that Herbert was looking at.

It thus became obvious that some aspects of Fellow's remarkable language understanding were not completely linguistic. In some instances, it was clear that the dog was responding first to his master's tone of voice, which suggested that he do something. The dog then picked up the location of the object that he was to do something with or to, by looking at Herbert's head and figuring out the direction to go in. Once the dog encountered an object, there were really only a few things that he could do. If the object was small, he could pick it up and bring it back. If the object was large, heavy, or fixed in place, he could simply look at it, or press his nose against it to indicate that this was the object in question. In some cases, where the object was furniture, or something solid of intermediate size, he could jump on or over it, or lay his head on it.

Let's now look back at the performance of that other German shepherd dog, London, under the command of Charles Eisenmann. Much of the performance had been filmed, and an edited segment of it was later shown on television. Knowing the "secret" of Fellow's terrific performance, I now kept my eyes on Eisenmann as he gave his commands. Sure enough, on the Spanish command to close the door, Eisenmann glanced in the direction of the door just as he finished saying, *"Cerra la puerta!"*

The broadcast segment also included the color vision test that had disturbed me so much. Now I could see that when Eisenmann said, "London, point to something in this room that is red," he was actually looking in the direction of the red mug on the little table next to the host. London followed that line of sight directly. Because I was now alerted to what was going on, it now became clear to me that London was simply pointing his nose at the edge of the little table. He didn't actually appear even to be looking at the cup. Apparently, the audience (including me) as human observers who could see the color red, quickly identified the red coffee mug and assumed that this was what London was pointing out. It may well have been that the dog was deliberately pressing his nose against the table itself, since it was the most obvious object in the direction that his master was looking. Similar slight head turns preceded the dog's identification of both the blue and yellow objects.

Please do not use these observations to assume some form of conscious fraud on the part of Eisenmann. The situation is similar to one encountered in the early 1900s, when a retired German mathematics professor, Herr von Osten, taught his beloved horse, Hans, lessons in history, mathematics, and spelling. In some cases, the horse would show its scholarly ability by answering multiple-choice questions. Thus, it might be asked which of four words was correctly spelled and then given a list such as (1) atc; (2) cta; (3) cat; and (4) tca. The horse would answer by tapping its foot an appropriate number of times (here, of course, it would be three taps). Van Osten was not interested in showing off the horse for money, but he invited small and select groups, often including well-known behavioral scientists, to view the horse and test him for themselves. All left convinced that the horse had surely earned his popular nickname, "Clever Hans," since he seemed to have extensive knowledge of language and also good mathematical, historical, and geographical knowledge. The true nature of Hans's gifts was revealed by the experimental psychologist Oskar Pfungst, who showed that the horse was taking his cues from unconscious, almost imperceptible shifts of head and body posture in the members of the audience. The major signal was the movements created by the involuntary relaxation of the tension in his observers when the hoof taps reached the correct number. This helped to explain the puzzling finding that Hans's accuracy in understanding language dropped as twilight approached, since in the darkness the horse could no longer see its audience's indications of the correct answer.

In the case of London and the other dogs, there was also a clear clue available. Glancing toward an object that you are talking about is a natural and unconscious act. Unfortunately, in this case where we are trying to understand the extent of a dog's ability to interpret spoken human language, these glances serve as an alternate form of communication. All that the dog needs to do is read the direction indicated, and the tone of voice which tells it to do something with an object in that direction. I think that at some level, Eisenmann may have been aware that there was something about his looking at the dog and the situation that was important, since he later wrote: "Although once in a while I give my dogs commands with my back turned, I frown upon the practice."[2]

Just how subtle are these orienting cues that we give to dogs when we are telling them where to go and what to do? It could be that dogs are responding to the fairly gross movements of our head and body, or it could be that they are actually following the direction in which our eyes are gazing. Human beings are quite good at picking up the direction that other humans are looking, so this is a possibility. Professor Warden tested this with Fellow, by having his master wear a blindfold. When Herbert was told to orient himself in a particular direction appropriate for a command, Fellow did fine—even though his master's eyes were no longer visible. This means that it is likely that the dog was using the larger orientation movements associated with head and body, rather than requiring an actual directional gaze from the eyes.

The important conclusion from all of this must be a realization that sometimes, when dogs appear to be responding to our spoken words, they may not be actually doing so. It is true that they can understand many words and sounds, but it is their subtle ability to pick up orientation clues from us that makes them appear to be even more knowledgeable. A set of words spoken to the dog might include sounds that it understands, such as "Take," "Bring," "Find," "Jump," or "Go to." However, other aspects of our spoken sounds may have no meaning at all. Thus, if I tell my dog, "Take the red thing," it is my spoken words indicating the action, "Take," that starts him moving to go bring me something, but it is the turning of my head and body that actually puts him on line with the red ball that I want him to retrieve for me. It is our all-too-human bias to think that the spoken language is the only thing that is important. My dog doesn't have a clue as to what "the red thing" is, but whatever is in my line of gaze—whether red, white, or green—is what he will bring back to me. While we often think that a dog is interpreting our language by listening, he may well be attending even more closely to what our body is doing. Dogs are masters at reading body language, even when we're not aware that we're communicating in that way. Later on, we will see that they can deliberately use their own bodies to send complex messages.

Up to now, we have only been considering how well dogs can interpret human language. Obviously, they are pretty good at interpreting the *meaning* of what we want them to know, whether that meaning is extracted from language or from other cues. Given our brief look at the receptive language of dogs, let's now turn to a consideration of their productive language in order to see what abilities dogs have in communicating with others.

5

Animal Noise or Animal Speech?

Perhaps the best way to begin to try to understand the language of dogs is to look at the way in which dogs speak. For human beings, speech, defined as the making of meaningful sounds, is so natural a way of communicating that many people think that speech is language. We have already seen that dogs will never be able to shape the complex types of sounds that humans create because of their physical limitations; however, dogs can, and do, make sounds, and these sounds are used for communication purposes.

Language in the form of sounds has certain advantages over most other types of communication. Let's compare sounds to visual signals. Visible body language is an important means of communication for dogs and has some survival benefits. A message in the form of a visible sign is silent, yet can be picked up from a distance. The location and source of the message is easily detected as well. The signal can be "turned on" or "turned off" instantaneously, and it also can be made to vary in intensity by making the signal movements more vigorous, faster, or larger. Complex information can be encoded in simple visual signals, such as tail wags or head movements, if the dog has already learned to translate these. This means that the language of visual signals is quite flexible, with no end of possibilities. Why, then, did sound signals develop?

For an animal, the problem with visual language is that its very advantages can be turned against the individuals using it. Since the signals must be seen to be received, it is not always possible to prevent a predator, or potential prey, from seeing the sender as well. The use of body language and visual signals requires highly developed eyes, with very good resolving acuity to perceive the finer details. Even if the receiver has good eyesight, visual messages will start to lose their details if the sender is too far away. Fog, smoke, or difficult viewing conditions also destroy the clarity of the message. Furthermore, any physical obstruction, such as trees, rocks, or walls, will block visual communication be-

45

cause these messages do not travel over or around obstacles. Visual communication also requires sufficient light in the environment for the signals to be seen. It would not seem sensible that evolution would cause an animal to rely completely on a communication channel that shuts off at night and cannot be turned back on until the light of dawn.

Sound signals seem to surmount many of the drawbacks associated with vision. Although sound communication requires a sensitive hearing apparatus, the sounds can travel long distances and still be interpretable. Sounds are not stopped by fog or total darkness, and they can go around corners, and through or over many obstacles such as forests, rocks, and walls. That, of course, is why it is so difficult to sneak up on a wild creature, even if it is asleep or dozing. The sound will reach the creature long before you see it. The flip side of that is that the sound signal can be sent from a hidden place, allowing the sender to remain concealed. The source of the sound can be accurately located by two keen ears, but it can also appear to come from a false or ambiguous location (as any good ventriloquist can demonstrate). High-pitched, brief sounds, such as a squeak, are difficult to localize. However, if another animal, say the squeaker's mother, has been listening, and already knows the sender's location, they can interpret the message and come at once.

When it comes to sounds, we certainly know that animals make a lot of them. It seems intuitively sensible to suggest that such sounds would not have evolved as part of the normal behavior patterns of animals if they did not have a specific function. A number of scientists have come to the conclusion that many of the sounds that animals make do seem to be similar to what we call "words" and to have very specific meanings that are understood by other members of the species. The most compelling evidence for this comes from monkeys (perhaps because people tend to be more impressed by what monkeys do than by what four-footed animals do). One such monkey is the vervet, from Africa. A slim, graceful monkey, with long arms and legs, and a fairly flat face, the vervet is sometimes known as a savanna monkey, since it spends most of its time on or near the savanna. It lives much of its life on the ground, foraging for fruit, leaves, and other vegetation. Vervets got their name from the French word *vert,* which means green, because the soft, dense fur on their backs has a greenish hue, which contrasts with their pale yellow or white underparts and black faces, hands, and feet. More important for our interests, they also have a distinctive vocabulary which warns others that possible predators are near. Not only do they sound the alarm when a predator is detected, but they also use specific "words" to tell the other members of their band what kind of predator they should look for.

Psychologists Dorothy Cheney and Robert Seyfarth[1] of the University of Pennsylvania conducted much of the language research involving vervets. They found that there are three main predators which hunt these handsome little monkeys: the leopard, the eagle, and the snake. A vervet monkey that detects a leopard produces a loud, barking sound in a rapid-fire burst. If it detects an eagle, it gives a different kind of bark that sounds like forced laughter. If a monkey spots a snake, it makes a high-pitched chattering sound. For the other vervets, each of these sounds acts as if it were a specific word or alarm signal, and each produces a different kind of behavior. When the bark that means "leopard" is heard, all of the monkeys stop their activities and run for the safety of the trees. When a vervet hears the chuckling sound that means "eagle," it looks up quickly to scan the sky before making a dash for the cover of low bushes. When a vervet makes chattering sound that means "snake," all the others stand up on their hind legs to search the ground around them for their low-slung and stealthy enemy.

To nail down the fact that these sounds were acting as words, these two researchers first had to rule out some other possibilities. The most important of these is that both the sounds and the specific protective behaviors that the monkeys exhibited were not meant to communicate, but were really just something like an emotional response caused by the sight of the predator. To test this, Cheney and Seyfarth made recordings of the various alarm calls. The recordings were later played back when no predator was in the vicinity. The result was that all of the monkeys that heard the recorded calls still responded in the appropriate way. This means that the monkeys were extracting the meanings from the sounds themselves and treating them as words.

There was another way in which vervet words seem to be similar to words in human language. When young children are learning to speak, they often make the mistake of using the same word for several similar things. Young vervets also make these kinds of mistakes. A young vervet might use the eagle alarm call when it sees a leaf falling, or the leopard alarm call when an antelope wanders by, or the snake alarm call when it spots a hanging vine. As they grow older, the vervets make many fewer mistakes, suggesting that they are learning and refining their language abilities. Along the way, they tend to do a lot of looking toward their mothers when another monkey sounds the alarm, as though learning what they have to do by watching the way that older, more experienced animals respond. They also seem to check to see whether other monkeys agree with their choice of an alarm call when they actually sound one. Over time, these young vervets become more precise speakers of "Vervish" by listening and observing other monkeys.

If the vervet sounds were a sort of primitive language, then, like all living languages, it should be capable of creating new words to describe new things or conditions in the environment. Modern human languages have had to create such words as "telephone," "computer," and "laser" as each of these things appeared on the scene and became important. Thus, we might ask, if a new kind of vervet predator appeared, would the vervets produce a new alarm call? Marc Houser, a psychologist and anthropologist from Harvard University, had an opportunity to observe this happening in the vervets. While he was passing through the same habitat that Cheney and Seyfarth had studied, he heard a chorus of alarm calls suggesting that a leopard was nearby. As he listened, though, he noticed that the calls were slightly off. He reported that "Rather than the intense, rapid-fire alarm calls typically associated with leopard sightings, these calls had a much slower temporal pattern of delivery, as would be perceived if the batteries of a tape recorder were run down during playback."[2] When he arrived on the scene, he found that the vervets were up in the trees and it was clear they had been signaling the presence of a lion.

During the course of studying vervets, lions had never before been seen attempting to hunt them. This is sensible, since the lion is slower than the leopard and is therefore less likely to be successful. Also, the small vervet is not much of a prize for the lion, who needs bigger game to provide adequate sustenance for itself and its pride. It seems that in this instance, because other sources of food had become scarce, the lion had turned to this non-traditional food source. The vervets responded by adding lions to the general category of hunting cats and produced a new "word" by modifying the sound of the usual leopard call. The environment had changed, and the vervet vocal language had responded by changing to include the new conditions, in much the way that human language expands to include new concepts.

This degree of language sophistication seems very complex, but one should not think that such linguistic ability is only possible in species with brains as complex as that found in a monkey. We find similar specific words, in the form of alarm calls, in many other animals. One example is the Belding's ground squirrel, which is a sociable creature that digs burrows. Because it spends most of the daylight hours in the open on rocks and around logs, this squirrel is quite vulnerable to aerial at-

tack by hawks and other birds of prey, and also susceptible to ground attacks by various mammalian predators, such as bobcats or badgers.
These two varieties of predators use very different hunting strategies,
with hawks relying on speed and bobcats relying on stealth. Like the
vervets, the ground squirrels have different alarm calls which specify
the predator that one of them has sighted. A high-pitched whistle is the
sound for "A hawk is approaching," while a rough chattering sound
means, "There's a mammal trying to sneak up on us." Just as in the case
of the vervets, these ground squirrels respond appropriately when they
hear one of these signals. They dash for cover when someone gives the
"hawk" call, and head for their burrows to stand watch when another
squirrel gives the "sneaky mammal" call.

Dogs are somewhere on the evolutionary scale between ground squirrels and vervets, so it would make sense to expect to find "meaningful"
vocalizations in canines. The sound most commonly attributed to the
domestic dog is barking, so it might be interesting to do some evolutionary speculation about how this barking came to be. In a later chapter, we will actually "translate" the meanings of various dog barks.

We will probably never have conclusive evidence to tell us how dogs
and humans first formed their personal and working relationship with
each other, but it is most likely the case that man did not initially choose
dog; rather, dogs chose man. Dogs were likely attracted to human campsites because humans, like dogs, were hunters, and animal remains, such
as bones, bits of skin, and other scraps of offal from the victims of recent
hunts, were likely to be scattered around human campsites. The ancestors of today's dogs (being ever food-conscious) learned that by hanging
around man's habitations, they could grab a quick bite to eat now and
then, without all the exertion involved in actual hunting.

Although primitive man may not have been very concerned with
cleanliness, health issues, or sanitation, it is still true that rotting food
stuff does smell, and attracts insects that will make humans uncomfortable. Thus, it is likely that dogs were initially tolerated around the
perimeter of the camps simply because they would dispose of the
garbage. This waste-disposal function continued for countless centuries
and is still being fulfilled by the pariah dogs in many less developed regions of the world. Anthropologists studying primitive tribes in the
South Pacific have noticed that on those islands where people keep dogs,
the villages and settlements are much more permanent. Villages without
dogs have to move every year or so, simply to escape the environmental
contamination caused by rotting refuse. This has even led to the suggestion that dogs may have been a vital element in the establishment of per-

manent cities in that bygone era before we learned the importance of public sanitation.

Once the wild canines that would eventually become dogs were attracted to human settlements, our ancestors noticed an added benefit. Remember, early humans lived in dangerous times. There were large animals around which looked on humans as potential sources of fresh meat. There were also other bands of humans with hostile intentions. Since the canines around the village began to look upon the area as their territory, whenever a strange human or some wild beast approached, the dogs would sound the alarm. This would alert the residents in time to rally some form of defense if needed. As long as dogs were present, the human guards did not need to be as vigilant, thus allowing for more rest and a better lifestyle.

It takes only a short journey to get from the dogs guarding the village to a personal house dog. Humans now knew that dogs would sound the alarm if their territory was invaded. Suppose that this idea was taken one step further. A dog which considered a home as its territory would then provide a personal warning for a family. This might serve the benign purpose of alerting the family to the approach of visitors (a sort of canine doorbell), or warn of the approach of someone with malicious intentions (a canine burglar alarm). Clearly, this was one of the motivations for taking puppies from the wild dogs, bringing them into the home, and domesticating them as house dogs.

When we say that the canines of early man barked, we are not talking about the kind of barking that dogs exhibit now. The sounds originally made by these ancient dogs were probably much more like those of the wild canines of today. Wolves, jackals, foxes, and coyotes rarely bark, and the noise that they make is far from impressive. I remember the first time I heard a pack of wolves barking when we approached their den. I easily recognized the sound as barking, but was surprised at how restrained it was. Domestic dogs set up a continuous string of machine-gun-like bursts of barking. The wolves' bark was much softer, sounding like a breathy "Woof." It never appeared in strings, but simply a single, monosyllabic bark, followed by a pause of two to five seconds, then perhaps another bark. Over a half-minute period I counted four modest barks, while a domestic dog may give thirty or more, very much louder barks in the same time period when warning of the approach of a stranger.

The earliest "dog owners" at some point must have noticed some variations in the amount of vocalization among the various canines that shared their camps. It seems obvious that for personal and community security purposes, the most effective dog is one with a loud and persistent bark. Apparently, humans began a sort of primitive selective breeding program to create such dogs. A dog that barked was kept and bred with others that also barked. One that did not bark was simply disposed of as being useless. This seems to explain the eventual divergence in vocal tendencies between dogs and wild canines.

To support the possibility of such a "controlled evolution theory," there is historical evidence of modern dogs being bred to bark. This involved the development of terriers. The terrier is a specialized hunting dog. The Latin root *terra* in "terrier" stands for earth or ground, and signifies the special ability of this type of dog, which is to follow game into its burrow or some natural crevice, and either flush it out or kill it. The earliest terriers barked, but no more so than any other domestic breed of dog. They barked when people approached their homes or their territory, and to sound a general alarm. However, dogs, like their wild kin, do not bark when hunting. This would be counterproductive. A bark would alert the game to the presence and exact location of its predator, and would make it easier to escape. Thus, most dogs hunt and attack in total silence.

Silent hunting may be a good thing in the wild, but when hunting animals in their dens or burrows for the benefit of humans, it is not so good. The sound of the dog underground is what tells the hunters where to dig to uncover the fox or badger, and also to retrieve their dog. Because terriers did not bark while pursuing or attacking, hunters designed special collars for these dogs, containing bells or clappers. The noise could then guide hunters in their chase and digging. Unfortunately, this was not a workable solution. Many dogs choked to death when their collars caught on some obstruction underground, others died because the hunters couldn't hear the tinkle of bells when fox and terrier were lost under the ground in their final confrontation. A barking dog, however, could be heard, and hence found, with no increase in jeopardy for the animal.

This led to the systematic breeding of terriers to bark. Dogs which readily barked when they were excited were singled out for special breeding with other terriers that readily barked. By the last quarter of the nineteenth century, virtually every dog that you might call a terrier was an excited barker. Size was not an issue. A tiny four-pound Yorkshire terrier who stands only 9 inches at the shoulder will bark more vigorously and with less provocation than will a 120-pound Great Dane, who stands 30 inches at the shoulder. This is not because the tiny terrier is either braver or more frightened, but rather because the terrier has been bred to bark.

Probably the most systematic scientific data obtained on the inheritance of barking in dogs comes from two psychologists, John Paul Scott and John L. Fuller, who for around fifteen years studied the genetics and behavior of dogs at a special facility, the Jackson Memorial Laboratory at Bar Harbor, Maine.[3] One of the dog breeds that they studied was the Basenji. The Basenji is a handsome African hound, with a pointed face, pricked ears, and a tail curled over its back. It is of medium size, weighing around 22 pounds and standing around 16 inches at the shoulder. One of the most striking characteristics of the Basenji is that it rarely barks. Much more often than barking, it will make a sort of soft howling sound which often reminds me of the sound of an Alpine yodeler, while at other times it makes a soft laughing sound. It can bark; it's just that unless the dog is very excited, it seldom does so. There has been some speculation as to why the Basenji rarely barks. The most common theory is that barking to alert the approach of a strange person or animal might be non-functional or even dangerous behavior in the African forests. Some naturalists have noted that leopards are fond of dog meat, and it may be that the dog that barks in that environment simply attracts attention to itself and thus makes it more likely that it will come to an untimely end.

Scott and Fuller set up a series of social dominance tests in order to see how various breeds would behave. One aspect of the behavior that they monitored was the amount of barking. The dominance test consisted of putting two dogs into a pen with one very desirable bone. Dogs put into such a situation will often bark to either threaten the other dog or try to lure that other dog away from the bone. In this situation, only about 20 percent of the Basenjis ever barked, as opposed to almost 68 percent of the cocker spaniels tested. Most Basenjis who did bark gave only one or two low, wolflike "woofs," with the noisiest of them giving twenty barks over the ten-minute test period. In comparison, 82 percent of the spaniels barked more than the noisiest Basenji. One cocker spaniel gave 907 barks in ten-minute period, which is more than 90 barks per minute! Thus, each Basenji is not only less likely to bark, but when it barks, it will make many fewer utterances.

The next step was to deliberately cross the silent Basenji with the noisier cocker spaniel. The result was a litter of dogs that was almost as likely to bark as the original group of cocker spaniels (60 percent as opposed to 68 percent). This suggests that the tendency to bark is not only genetic in domestic dogs but is probably a dominant gene. It means that a dog with both a barking and a barkless gene will turn out to be a barker, which probably explains why it was so easy for early dog breeders to create a domestic dog that is likely to bark. Once the trait is found, it is easy to cultivate and to maintain since it overrides other more silent tendencies.

The dominant tendency to bark, however, is not the whole story. Although the Basenji and cocker spaniel mixes are almost as likely to bark as a purebred cocker spaniel, the number of barks that they give during a period of time is reduced. Thus, for the purebred cockers, 82 percent barked more than twenty times during the test; for the Basenji/cocker mixes, only 49 percent did. This shows that there are two different inherited tendencies. The first is the likelihood that a dog will bark (or the amount of stimulation needed to get it to bark), while the second is the actual amount or pattern of barking that it will produce, once it has started barking.

We have discussed a theory about the evolution of barking in modern dogs but as yet have not discussed whether different barking sounds can be interpreted as having different meanings (as in the predator calls of the vervet and ground squirrel). To do this, we will first have to learn how to listen to the sounds that dogs make and to distinguish how various barks and other vocalizations can differ. The truth of the matter is that most people don't seem to pay much attention to the shades of difference in the various dog barks. Because of this, we are losing much of the content of the messages that dogs are trying to send to us. An example of how insensitive we are to a dog's voice is the fact that to the human ear there is not even any consensus as to what the basic sound that dogs make is. To the English or American speaker, the dog says, *bow-wow, woof-woof,* or *arf-arf.* To the Spanish, it is *jau-jau,* to the Dutch, *waf-waf,* to the French, *woa-woa,* to the Russian, *gav-gav,* to the Hebrew speaker, *hav-hav,* to the German, *wau-wau,* to the Czech, *haff-haff,* to the Korean, *mung-mung,* and to the Chinese, *wung-wung.* Could it be that our dogs are speaking in different languages? It seems more likely that we are just not very good listeners.

6

The Dog Speaks

For human beings, language sounds are fairly arbitrary. There is no set of words that have a common meaning for all members of our species. Many different sounds, in different languages, can mean the same thing. The sounds associated with the words *perro, chien, Hund,* and "dog" all mean the same thing, although there is virtually nothing in common among the sound patterns which make up these words. Humans have sometimes tried to eliminate the problems created by the multiplicity of human languages by creating a "universal language." The best known of these is Esperanto, but unfortunately it has had minimal effect on human language diversity. The sounds that animals use to communicate with each other, however, do have much more uniformity. These sounds are different for different species, but (except for certain regional "dialects" among birds) within any one animal type, there seems to be some sort of fairly common or universal language.

The universal language code of animals, what we might call Evolutionary Esperanto, includes a number of common sound patterns that are used for communication. These have been shaped, not by scholars and linguists but rather by the evolutionary pressures operating on the sounds that animals typically make. Evolutionary Esperanto allows not only for the various groups of dogs to understand each other's vocal signals, but also for other species (including humans) to extract a good deal of the meaning from these signals. In the Evolutionary Esperanto of animals, the basic rules for understanding have to do with three dimensions: the pitch of the sound; the duration of the sound; and the frequency or repetition rate of the sounds.

Consider the meaning of varying the pitch of a sound. For growls, barks, and other sounds, low-pitched sounds (such as a dog's growl) usually indicate threats, anger, and the possibility of aggression. Low-pitched sounds basically mean, "Stay away from me." In general, high-pitched sounds mean the opposite. They make the statement, "It's safe to come closer," or ask the question, "May I come closer to you?"

The naturalist Eugene Morton,[1] working with J. Pope at the National Zoological Park in Washington, D.C., analyzed the sounds of fifty-six species of birds and mammals and found that this Law of Pitch held for all of them. In the same way that dogs growl, so do elephants, rats, opossums, pelicans, and chickadees. All of these growls seem to mean, "I don't like this," or, "Keep your distance," or, "Beware." In the same way that dogs whimper or whine, so do rhinoceroses, guinea pigs, mallard ducks, and even wombats, and these whimpers have the general meaning of "I'm no threat," "I hurt," or "I want." Psychologists have demonstrated these same characteristics in human speech. When angry, or threatening, the pitch of a human voice tends to drop to a lower register. On the other hand, when inviting someone to come closer and be friendly, the human voice tends to rise in pitch.

The question is, why should dogs, or elephants, or pheasants, or you, use and understand this Law of Pitch? The answer begins with the simple observation that big things make low sounds. Take two empty water tumblers, one large and one small, and tap each with a spoon. The large one gives a lower-pitched ringing sound. A longer string on a harp or piano gives a lower sound, as does a longer organ pipe. The physics of resonance apply to animals as well as inanimate sounding objects. Bigger animals make lower-pitched sounds than smaller animals. It is not that big animals decided to make lower-pitched sounds in order to let other animals around them know that they were big; the reason is basic physics. Nonetheless, because evolution works through survival, it was the animals that learned to avoid things that made low-pitched sounds who were also more likely to avoid fatal encounters. The flip side of the coin is that animals are more likely to survive if they learn not to run from high-pitched peeps and whimpers, since these are most likely made by small, non-threatening creatures, while running away from them in panic might cause injury or attract the attention of something larger and more dangerous.

Here is where evolution and the development of communication begin to weave their magic. Suppose that you are an animal who is sending signals to those around you. Since you know that other animals are paying attention to the pitch of your signals, you can now deliberately use that as a means of communication. If you want to make another animal move away, or stay out of your territory, you could send a lower-pitched signal, like a growl, suggesting that you are larger and more dangerous. Conversely, you could also use a high-pitched signal, like a whimper, to suggest that you are rather small, and therefore it is quite safe to approach you. Even if you are large, if you wish to signal that you intended no threat or harm when you approached another animal, you could indicate that you intended to act like a small, harmless creature by whimpering or whining.

This kind of behavior is overtly manipulative, of course. You do not change your size simply by changing the pitch of your sound signals. So, why would the receiver of this signal respond to your pitch variations at all, since they no longer represent the physical reality? The reason is that there is a survival advantage for the animal receiving the signal to respond to the pitch. An animal that is growling in a low-pitched voice is definitely one that it pays to avoid. You are better off staying away from the animal making these signals, whether it is actually large or whether it is just a small animal in a hostile mood ready to attack. Little teeth in an angry mouth can cause large wounds. An animal making high-pitched whining sounds is another thing altogether. There is no need to avoid the source of these high-pitched sounds, regardless of whether the animal is actually small or whether it is just feeling friendly and sociable, since it is signaling the exact opposite of a threat.

Evolutionary behaviorists would say that these pitch-related signals have now become "ritualized." Our responses to these signals have become independent of the physical facts which made them adaptive in the first place. These ritualized signals will now serve a very useful language function, and they can also reduce the amount of unnecessary violence and aggression among animals that live in social groups. If a wolf approaches the pack leader and is met with a low growl, he can assume that the emotional state of the leader is angry and aggressive. He can now avoid trouble by simply staying away, before teeth are bared and bodies damaged. Alternatively, the approaching wolf can begin to make high-pitched whimpering sounds to indicate that he is not threatening or challenging. Under these circumstances, the leader might well stop growling and allow him to approach. In either case, blood is not spilled because both know the meanings of the ritualized signals. These meanings have evolved because they carry useful information and they work to maintain social harmony.

It is important to note that growling only works in the sense that it is a ritualized signal which is attempting to change another individual's behavior. Specifically, the growl is meant to warn someone away. A dog that has decided to attack does not growl—it simply attacks. If a dog has been growling and the target of its communication does not begin to

back off, the growling may well stop. This doesn't mean that the hostility has ended. It may well be that the dog recognizes that its warning was
not accepted and that the only alternative left is to fight. Now silent, the
dog lowers its head slightly; its curled lip may tremble, and then with no
further sound, there will be a lunge and a snap. Dogs that have decided
to attack do not give vocal signals. If you have ever seen a demonstration
of a police dog working, when the signal to attack the person pretending
to be a fleeing criminal is given, the dog simply races out in ghostly silence and clamps its jaws on the criminal's padded arm. Once the fight
has started, growling may return, giving the opponent the signal to
break off the battle and move away. A frightened dog that decides that
the only way to save itself from harm is to flee will be equally silent. A
dog dashing away, trying to put as much distance as possible between itself and some perceived threat, often makes no sound at all. The situation has clearly passed beyond the stage of social communication. In
both the case of fear and the case of anger, when the sounds no longer
serve as language, they stop. Why continue to make sounds if they are
no longer useful signals that may affect the behavior of some listener?

The next important feature of Evolutionary Esperanto is the duration of the sound. This can be used to modify the meanings of sounds of
particular pitches. The way that sound duration combines with pitch is
a bit complex. Basically, shorter sound durations are associated with
higher-intensity fear, pain, or need. Take the high-pitched whining sound
of a dog as an example. If you shorten its duration, it becomes a yelping
sound, which may mean that the dog has just experienced a painful
event, or is terrified and about to run for its life. If you make that sound
longer in duration, it becomes a whimper, which can mean pleasure,
playfulness, or invitation. Generally speaking, the longer the sound, the
more likely that the dog is making a conscious decision about the nature
of the signal and the behaviors that are about to follow. Thus, the threatening growl of a dominant dog that has every intention of holding his
ground and not backing down will not only be low-pitched but also will
be long and sustained. If the growl is in shorter bursts, and held only
briefly, it indicates there is an element of fear present and the dog is worried about whether it can successfully get away with a full attack.

The third dimension in the Evolutionary Esperanto of dog vocalizations is the repetition rate of the sound. Sounds that are repeated often,
at a fast rate, indicate a degree of excitement and urgency. Sounds that
are spaced out, or not repeated, usually indicate a lower level of excitement or a passing state of mind. A dog giving an occasional bark or two
at the window is showing only mild interest in something. A dog bark-

ing in multiple bursts and repeating them many times a minute at that same window is showing a much higher level of arousal. It is signaling that it feels the situation is important and perhaps even a potential crisis.

We can see how these dimensions interact if we consider what dogs mean when they use various vocalizations such as barks, growls, howls, whines, and so on.

BARKS

While Eugene Morton was analyzing animal sounds to come up with the Law of Pitch, he learned that many species bark, in the same way that many whine and growl. A squirrel, a monkey, and even a rhinoceros may bark. Even the cheeping sound of some birds follows the basic pattern of a bark. If you tape-record some of these bird sounds and play them back at a slower speed, they sound remarkably like a dog's bark.

Originally, barks probably were simple alarm calls to announce that someone was coming, much like the medieval trumpet that announced that people were approaching the gates of a fortress. Notice that the alarm does not yet tell us whether the arrivals are friendly or hostile; it just alerts us so that we can prepare for action. That is why a dog may bark just as loudly when its master starts walking up the steps of the house as when it feels there is a burglar trying to gain entrance.

The bark also serves the same function as the challenge from a sentry. The "Halt, who goes there?" that the sentry gives announces the individual's presence and allows the guard to gather a bit more information. As in the case of a sentry, once the new arrival has been identified, the behaviors will change. In dogs, the barking stops and the animal may substitute whimpering and tail-wagging as a friendly greeting for a familiar person. It could also stop barking but begin growling and threatening an attack if the visitor is perceived as hostile.

When you analyze the sounds of the bark, you find that they are composed of a changing pitch that rises and falls sharply. In this sense, they combine the rough qualities of a growl with the tonal qualities of a whine. Since barks are in the middle of the pitch range, it is easy for the dog to make the pitch slightly higher or lower in pitch, in order to produce many shades of meaning. Let's look at some of the basic bark patterns and their interpretation.

Barking in rapid strings of three or four with pauses in between, midrange pitch: This is a fairly indefinite alerting call. It means that the dog senses something is there, but it is not yet identified or close enough to actually be a threat. It is really a suggestion that other pack members should begin to gather at this location. Thus, this sound might mean, "I suspect that there may be a problem or an intruder near our territory. I think the leader of the pack should look into it."

Rapid barking, midrange pitch: This is the basic alarm bark. It means, "Call the pack! Get ready for action! Someone is coming into our territory!" Since there is now a high rate of barking, this means that the dog is aroused and feels that the visitor (or problem) is much nearer.

Barking still continuous but a bit slower and lower pitch: The lowering of the pitch and the slowing of the rate suggests that the dog is sensing an imminent problem. Thus, this sound means, "The intruder [or danger] is very close. I don't think he's friendly. Get ready to defend yourself!"

Barks have extended their meaning well beyond their original alarm call function. By adding certain sound nuances, they have become ritualized signals that convey some more subtle meanings. For instance:

A prolonged string of barks, with moderate to long intervals between each one: These sounds are really in the form of "Wuff"—pause—"Wuff"—pause—"Wuff," and so forth. They are best understood as meaning, "Is there anybody there? I'm lonely and need companionship." This is most often the response to confinement or being left alone for long periods of time. Since dogs are social animals, separation from their pack can be very stressful. If the stress level is great enough, the pitch of the bark will rise out of the usual midrange, and will sound something like a yelp that has been subtly mixed with the bark. As in most cases, rising tones invite others to approach, so this kind of plaintive barking is also saying, "I'm still here. Have you forgotten about me? Please respond."

One or two sharp short barks of high or midrange pitch: This is the most typical greeting sound, and it usually replaces the alarm barks when the visitor is recognized as friendly. Many people are greeted in this way when they walk in the door. It really means, "Hello there!" and is usually followed by the dog's typical greeting ritual.

Single sharp short bark, lower midrange pitch: You will often hear this sound when a mother dog is disciplining her puppies. It's the same sound that you'll hear when a dog is disturbed from sleep or its hair is pulled during grooming, and so forth. The slightly lowered pitch is always associated with the idea of threat or annoyance, so this sound can be interpreted as meaning, "Stop that!" or, "Back off!"

Notice how certain small nuances and subtleties in the dog's verbalizations can change the intended meanings quite a bit. This is very similar to the way that changes in the inflection or tone of voice can change the meaning of statements in human languages. Thus, we can use the same two-word phrase, "It's ready," as either a statement of fact meaning, "It is now ready," or a question meaning, "Is it ready now?" We change the statement into the question by having the pitch of our voice go up at the end of the phrase for the question, rather than down as it typically does in statements. Inflections can drastically change the meanings of words. We all know that we can express agreement with something a person has said with the words, "Yes. Sure." However, by adopting an ironic tone of voice, these same words can be used to convey the message: "I don't believe a word you're saying."

Dogs use similar changes of inflection, which affect the duration and pitch of the sound to vary the meaning of a bark or other vocalization. You find these inflection changes particularly in the single or short bark sequences. Let's see how these nuance shifts work for dog communication.

Single sharp short bark, higher midrange pitch: This is the sound of surprise. You also can get this if the dog is startled in some way, and it means, "What's this?" or perhaps simply, "Huh?"

If this same single sharp short midrange pitch bark is repeated two or three times with a moderate interval between repetitions, the meaning changes to "Come look at this!", alerting nearby individuals to a novel event. The fact that the pitch is neither obviously high or low means that this utterance is driven by curiosity or interest, and is not, for the moment at least, signaling either fear or defensive anger.

If we modify the same type of bark so that it is not quite as short and sharp, but has a more deliberate delivery, the meaning has changed to "Come here!" This is more of an insistent command then in the previous case, and many dogs will use this kind of bark near dinnertime, when they are standing near their food bowl, to try to hasten your arrival with something to eat.

Lowering the pitch to a relaxed midrange, just enough so that the bark still sounds short but seems to have lost any sharpness in the tone, converts it to "Terrific!" or some other similar expletive, such as "Oh, great!" I once had a cairn terrier who loved to jump, and he would give this single bark of joy when sent over the high jump. Unfortunately, since dogs are supposed to work silently in obedience competition, each of those barks would cost me a point. I never tried to do anything about this happy bark, since hearing it gave me the pleasure of knowing that my dog was having fun. His joy in working was more valuable to me than an extra point or two in my score. Other dogs give this same bark when you give them their food dish or pick up their leash for a walk.

Certain dog sounds are associated with particular activities. One of the differences between domestic dogs and wild canines is that we have preserved a number of puppylike characteristics in domestic adult dogs, which are lost when the wild canine becomes an adult. A delightful example of these puppy behaviors is the desire to play, which has been maintained in full strength in all dogs. Our dogs have also developed a set of vocalizations or specialized barks that invite other dogs to play, or express the pleasure they are having when they are engaged in play.

Stutter-bark, midrange pitch: Imagine that we could transliterate the sound of a dog's bark as "Ruff." The stutter-bark would then transliterate as "arr-Ruff." It means, "Let's play!" and is used to initiate playing behavior. It is usually accompanied by a particular play invitation posture, in which the dog lowers its front so that its front legs are almost flat on the ground to its elbows, while its hindquarters are held high with the tail straight up. The stutter-bark may be immediately followed by the dog dashing to one side or another and then assuming the play pose again and giving another stutter-bark.

Rising bark: This is a bit harder to describe, although once you've heard it, the sound is unmistakable. It is usually a series of barks, each of which starts in the midrange, but rises sharply in pitch. It is almost a bark-yelp, but not quite that high-pitched or strident. This is another sound that is associated with play. It is not an invitation to play, but occurs during play, especially in rough-and-tumble games. It shows excitement and translates to "This is fun!" A version of this bark can be heard when some dogs get very excited at the prospect of their owner throwing a ball or Frisbee.

GROWLS

Although we have the image that growls are sounds associated with the great large and dangerous predators, such as tigers, lions, and bears, growling is actually a sound that is made by many other animals. Meek animals, such as the ring-necked pheasant, the opossum, and even some rabbits, growl. The purpose of the growl is to keep other animals away. Growls can be stand-alone "words," or they can be used to modify barking sounds to add a degree of threat.

Soft, low-pitched growling that seems to come from the chest: This is the classic growl from a confident and dominant animal, which means, "Beware!" or, "Back off!" It is deliberately used as a threat. Others who hear this sound usually respond by moving away and giving the dog a wider space. Failing to respond could trigger an attack. If a dog has been making this sound, and abruptly stops while not changing its posture to a more relaxed one, watch out. It may mean the dog has decided that "talking" is not working and the only recourse is to actual physical violence. Remember, most attacks begin in silence.

Soft growling that is not so low-pitched and seems more obviously to come from the mouth: Many people would describe this as more of a snarl than a growl, and, in fact, there may be a bit more lip curling associated with this sound. It has similar meaning to the previous growl in the sense that it means, "Stay away!" or, "Keep your distance!" The slightly higher pitch shows that the sound is made by a somewhat less confident animal that would really rather not fight but is willing to do battle if challenged.

Low-pitched growl-bark: This is a clear growl that leads to a bark. It sort of sounds like "grrrrrr-Ruff." By adding the bark, we are of course adding a component with a bit higher pitch. Remember that a higher pitch is more apt to be associated with less dominance and aggression. By combining the bark with the growl, the dog is calling for reinforcements. Thus, this vocalization means, "I'm upset and I'm ready to fight, but I may need some help over here." It is still a clear warning signal to keep away. This dog may prefer to receive some assistance from its packmates at this moment, but if you press it, it will act aggressively on its own.

Higher midrange-pitched growl-bark: The rising pitch added to the growl-bark is a sign that this dog is considerably less confident. It is saying, "I'm worried or frightened but I will defend myself." Although its confidence is thin, this threat is real, and the animal is very likely to fight back if pushed.

Undulating growl: This is a growl that changes pitch and goes from low midrange to high midrange. It may be broken into phrases, which are short growls with varying pitch, and occasionally there may be a kind of a semi-bark added as the pitch rises. The sound means, "I'm terrified. If you come at me, I may fight or I may run." This is the fearful-aggressive sound of very unsure dog. The conflict as to whether to stay and fight versus running for its life shows up in the pitch variations and occasional pauses in the growling.

Noisy growl, medium and higher pitch, with teeth hidden from view: Until you know a dog well, this growl might be a bit more difficult to interpret by sound alone. It sounds like a growl, but there is no low rumbling component. Often the best way to understand a sound is to consider how the communication fits into the larger pattern. In this case, the growling is audible but the teeth are not visible nor are the lips curled in a snarl. Actually, what this means is, "This is a good game!", "I'm having fun!" It is usually part of the play sequence and may be tucked in between a series of stutter-barks. It usually indicates intense concentration, as in a tug-of-war, a chase after a stick held by another dog, or play-acting aggression. If you are quite familiar with a particular dog, this growl will be quite distinctive, even if you are not consciously analyzing it. In a familiar dog you might say to yourself something like "This isn't convincing. He doesn't mean it as a threatening growl."

HOWLS, YOWLS, AND BAYING

While domestic dogs bark more than wolves or their other wild canine cousins, our dogs actually howl a lot less. In wolves, the howl has several functions. One of them is to assemble the pack for hunting. Since wolves hunt early in the evening and early in the morning, it is not surprising that those are the times when we are most likely to hear wolves howling. The howls assemble the group, which may have dispersed into the underbrush to sleep through the evening or to rest out of sight during the day. Since our domestic dogs have a food supply that is presented to them by their masters, they do not have call together their pack so that they can have a synchronized hunt.

Another purpose of the howl is to reinforce the identity of the group. Upon hearing the howling, group members gather together and join in the song of their pack. Because of this, dogs often howl when they are forcibly shut away on their own, or otherwise isolated from their family and pack. This howl of loneliness has the same function as the group howl: it is an attempt to attract other dogs.

Not all howls are the same, however.

Yip-howl: This sounds something like "yip-yip-yip-howl," with the final howl quite prolonged. It usually means, "I'm lonely," "I feel abandoned!", "Is there anybody there?" It is the howl you will most likely get from a dog that has been removed from the company of its family, perhaps locked away in a basement or garage for the night.

Howling: The traditional howl, which starts without any fanfare and produces a continuous, prolonged sound. It may occasionally begin with a slightly higher pitch before moving to the main tone, and sometimes the pitch may lower a bit toward the end. It has a more sonorous sound to the human ear than does the yip howl, and is often described as "mournful." This vocalization says, "I'm here!" or, "This is my territory!" A confident animal will often howl simply to announce its presence. This howl is often given in response to a yip-howl from another dog, when it can mean, "I hear you out there!"

Other animals may join in the chorus. Once the howling begins, it often turns into a joyous celebration—the dogs, or wolves, happily announcing their own presence and their camaraderie with others of their species in what might be termed a spontaneous canine jam session. This vocal performance may go on for quite a while and involve animals from all over a region or neighborhood. It is during such a wild concert that canines show their musical sensitivity. Recordings of wolves have shown that a howling wolf will change its tone when others join the chorus. No wolf seems to want to end up on the same note as any other in the choir. I have often felt that humans who fail to respond to the howling of a dog by joining into the chorus are in some ways derelict in their duties as pack members. My wife doesn't agree, however, especially when it is possible that the neighbors might hear me howling and draw the wrong conclusions.

Sometimes, dogs howl during a musical performance. This most often happens when the instruments are wind instruments, particularly reed instruments, such as clarinets or saxophones, or flutes. Sometimes they can be induced to howl by a long note on the violin or even by a human holding a long note while singing. Presumably, these must sound like proper howls to the listening dog and it must feel the need to answer.

Bark-howl: This is one of the sadder sounds that a dog can make. It begins with two or three barks and ends with a howl, and the sequence may be repeated several times. It is usually made by an animal that has been relatively isolated, for example, locked away all day in a yard with no access to human or other companionship, and is triggered when a stranger or another dog approaches the place where the dog is being kept. The bark shows the animal's desire to bring his pack members to him, just in case there is a problem, while the howl indicates the expectation that no one will respond. In effect, this sound means, "I'm worried and alone. Why doesn't somebody come and help me?"

Since howling is an attempt to gather the pack and to avoid loneliness, it may explain the long-held superstition that howling dogs are a warning that someone in the family is about to die, or some disaster is about to occur. This is usually associated with the belief that dogs have some sort of mystical powers that allow them to foresee the future. It is often the case that things that appear to be connected with bad events are themselves believed to be evil, however, our faith in dogs seems to have caused us to interpret these howls that preceded a death as an attempt by our faithful friend to sound a warning to its human family so that they might know the danger was near.

If we ignore supernatural explanations, there is a simple alternate possibility to explain this association. Suppose that someone in a house is ill. Because of the need to care for that person, a dog who normally stayed inside the home might be viewed as a distraction, a bit of a bother at the moment, or a source of noise that could disturb the patient. For this reason, the dog might be put outside or shut away for a while. Thus, a dog who is normally surrounded by his family, and might even usually sleep in the same room with the sick person, now finds himself alone. This loneliness might easily cause him to howl. Since there is already someone sick in the home, the likelihood that there will be a death is much higher than usual. Thus, people will only remember things like "Grandfather's dog never howled before, but the night that Grandfather died, he howled so mournfully because he knew the end was near." The truth of the matter is he never howled before because he was never locked up and isolated from his family before. On that night, since Grandfather was so sick, the family felt it best to shut the dog away. From such chance associations legends may grow. Of course, if we need a plot for another episode of *The X-Files* or *The Twilight Zone,* we should stick with the more mystical explanations.

Baying: Howling is quite different from baying, the sound that hounds make when tracking. When you first hear these two sounds, there is a vague similarity, but baying is more melodious. Baying contains many tone variations rather than holding a single tone for a long duration. To me, baying sounds something like the combination of a howl and a yodel. It is certainly a much more excited sound and is often filled with a happy enthusiasm.

Hounds bay to indicate that they have the scent of their quarry. It has some of the same "gather around me" intention that howling does, only the cause is not loneliness but rather cooperation on the hunt. Since at any one time only a few dogs in a pack may have the scent, the baying sound is interpreted by the other members of the pack as meaning, "Follow me! I've got the scent." As the scent becomes stronger, suggesting that the pack is now very close to its prey, the baying becomes a bit less melodious, since the individual sound phrases become shorter in duration, but more frequent, and the message now shifts to mean, "Let's get him!" or, "All together now!"

WHINES, SQUEAKS, AND WHIMPERS

The highest-pitched sounds that dogs make are interpreted by humans as whines or whimpers. The pitch gives away part of their meaning, in that they are designed to bring the listener closer to the whimperer and to show either fear or submission on the part of the whiner. These are also the sounds that puppies make, which is perhaps the reason why they can be used as pacifying and soliciting sounds. They suggest an absence of threat. They also suggest dependency and need.

Behavioral scientists have found that these short, high-pitched sounds are, in some ways, very special. They are pretty much the same sounds made by the young of most land-living vertebrates, whether wolf, bear, cat, alligator, chicken, or duck. They have two important qualities. First of all, they are very easy to hear and to separate from other sounds in the environment. Second, it is difficult to locate the exact source of such sounds. Both of these qualities are extremely important in the communication between a mother and her young offspring. She must, obviously, be able to hear any distress calls that they make. In addition, these distress calls must not give away the hiding place of her litter to any potential predators. The fact that the location is not available in the sound is not important to the mother, since she knows where she left her babies.

In puppies, the linguistic code is really quite simple. The louder and more frequent the whine or whimper, the greater the intensity of feeling behind the communication. In puppies, whining is an attempt to communicate a desire. This can be for food, for social interaction, or for play. It can also be triggered by bodily sensations, such as a full bladder. If the whining is ignored, it will become more intense and more frequent, before the puppy finally accepts the fact that no one is going to respond to it now.

The distressed or plaintive "I want" or "I need" whine rises in pitch toward the end of the sound. It can reach such high-frequency combinations that it sounds like chalk screeching down a blackboard, and can produce the same unpleasant physical and psychological feelings of discomfort as that mechanical sound. At its loudest and most demanding, it may sound much like a combination whimper, bark, and yelp, all in one complex utterance. The sound is sufficiently unpleasant that it becomes difficult to ignore and is certainly impossible to sleep through, which makes this particular sound quite useful as a means of attracting attention.

We can contrast this to the excitement whine mentioned earlier. The excitement sound comes in regularly spaced pattern, with only a few seconds between each sound. This whine drops in pitch at the end or simply fades out with no pitch change, so that it doesn't produce the discomfort in the listener that the distress whine does. Furthermore, there is a particular bit of body language that accompanies the sound. The dog looks at its master and then may twirl around in a little dance, perhaps looking alternately at its master's face and then the door if it is anticipating a walk, or perhaps looking at the face and then the food cupboard or food dish if it is anticipating dinner. My retriever Odin will often make these sounds while first gazing at me, then shifting his gaze to the shelf where I leave his Frisbee, then bringing his gaze back to me. The meaning is just as obvious as if he had said, "Come on now, grab that Frisbee and let's go out and play."

Adult dogs also use these sounds in special circumstances. In effect, when adults use these "puppy words," they are making themselves very small and puppylike in the presence of a dominant or threatening animal. These childish sounds tell the listener, "I'm small and weak and not a threat." They can also be a plea for help in times of dire need.

Soft whimpering: This is one of the most heart-rending sounds a dog can make. It means, "I hurt!" or, "I'm scared." The average person is most likely to hear this at a veterinarian's office, when the dog is suffering, or when a submissive dog is in a strange place that appears threatening. It is often accompanied by the dog averting its eyes to avoid direct contact with those around it, which reinforces the fact that this is a submissive communication. It is surprising to hear how close this adult sound is to the plaintive mewing that young puppies make when cold, hungry, or distressed. The idea that an adult from a predatory species is now making distressed puppy sounds is a clear indication of the physical or psychological ordeal the dog is undergoing.

Moan or moan-yodel: This sounds something like "Yowel-wowel-owel-wowel." It is clearly lower-pitched than whines and whimpers, but still in the mid- to higher end of the pitch range. It is an anticipation sound, which might be seen as resulting from spontaneous pleasure and excitement. It can be interpreted as meaning, "I'm excited!" or, "Let's go!" and usually is given when something that the dog really likes is about to happen. For most dogs, this has the same general meaning as the excitement whining we discussed a moment before. For reasons that are not clear, some dogs adopt an alternate sound to mean exactly this same emotion. I call this a "howl-yawn." From the name, you can derive the fact that this is a kind of sort of a howl (a bit higher-pitched than the typical howl) mixed with a sound something like a yawn that comes out as a breathy "Hooooooo-ah-hooooo." I have no idea why some individual dogs choose the moan-yodel, while others choose the howl-yawn or even the excitement whine to express this same hopeful expectation that something good is about to happen.

Yelps are different from barks in that they are higher-pitched and contain some of the elements of a whine combined with a bark. Most humans find a dog's yelping sounds to be a bit disturbing or distressing, probably because we pick up on the fear or pain component in them.

Single yelp or very short, high-pitched bark: This is the equivalent of a human shouting, "Ouch!" (or some short profanity) in response to a sudden, unexpected pain. A mother dog may actually punish young puppies who make their littermates yelp. A yelp from a too-hard bite during play usually ends the play sessions for puppies, and this is one of the ways that puppies learn to ease or inhibit their bite strength during play or other interactions with members of their pack or litter.

Series of yelps: A very obvious signal which either means, "I'm hurting!" or, "I'm really scared," this is the response to severe fear and pain. You can also hear a series of rapidly repeated yelps when a dog is running away after a fight, a severe threat, or a painful encounter. Under these circumstances, other dogs involved in the incident will usually not follow or pursue it. Apparently, this kind of yelp in such situations is read as a sign of surrender. The sound usually cuts off further aggressive action, since if the surrender of the yelping dog is accepted, no further aggression is needed to make the point.

SCREAMING

This sound seems much like that of a child in extreme pain and panic mixed with what I can only describe as a really prolonged yelp. The sound bursts are several seconds in length, and repeated. It is a sound of a dog in the most extreme pain, fearful for its very life. The distress carried by such a sound is unmistakable, although the species making the sound may be less clear. I have only heard this screaming a few times in my life; but once, when hearing it, I mistook it for the sound of a human child in some critical circumstances and charged across a construction site to provide assistance. The sound conveys such anguish and agony that I would be perfectly happy never to hear it again.

In well-established packs in the wild, or in well-bonded families containing more than one dog in the household, screaming will usually cause the other dogs to rally to the side of their hurt or frightened comrade. Their approach, however, is not done with the bravado that they use to support a dog barking out an alarm. Instead, they approach cautiously, just in case there is another predator around who is making their packmate scream and who also might be a danger to others in the vicinity.

Although these screams can be interpreted by the dog's own pack as a cry for help, such sounds can be a fatal mistake in the presence of unfamiliar dogs. Since this is the cry of an animal that is fearing death, it may produce a predatory reaction in another dog that doesn't know the screamer. This may cause the stranger actually to attack the dog making the sounds. This is not a sign of evil on the part of the attacker. You must remember that dogs are hunters. To such a predatory carnivore, screaming is the sound of a wounded animal. The indication of prey that is wounded and vulnerable can trigger a fast, forceful attack, which, in the wild, would reasonably ensure an easy next meal. The fact that these screams do not carry with them the immediate recognition that another dog is making the sound probably contributes to the likelihood of attack.

I saw this very thing happen once, outside a dog show. A man was walking a Malinois (which is a short-haired Belgian sheepdog that looks somewhat similar to a German shepherd dog) on leash toward the building housing the show. A van pulled up nearby in the parking lot and the driver opened the rear door. Before he could do anything else, a handsome white Samoyed jumped out. Unfortunately, the dog landed exactly on the place where someone had left a broken bottle. The sharp glass tore into the Samoyed's paws as it hit the ground, and it began to scream. The Malinois, which had been quite placid up to now and had shown no aggression toward any of the other dogs it had passed, suddenly leaped forward, jerking the leash out of its owner's hand. By the time the dogs were separated, the Samoyed was not only bleeding from the glass cuts but also in several places from the other dog's bites. The screaming of an unknown animal had simply activated the genetic code which causes a predator to attack.

Humans should learn to be alert for this screaming sound, since it is an important signal when dogs are having some sort of dispute. Generally speaking, I advise people never to interfere when two dogs are having a conflict. There are "rules" and "rituals" that dogs use to determine questions of social dominance, territory, and ownership of items. Fights generally proceed according to these rules, and there is seldom any bloodshed other than an occasional ear bite or similar minor wound. If dogs are baring their teeth and producing their fighting roar (a loud, continuous growl, occasionally punctuated by something that sounds like a person yelling, "Hey!"), we have a normal dispute. Usually, if the dogs are left alone to work things out, the conflict will end quickly without violence. It normally ends when one of the dogs backs off and shows submission. At that point, the incident is over. Although it is rare, it sometimes happens that a poorly bred dog will refuse the submissive gestures of the other and will not honor this signal to stop the conflict.

He continues the fight, and now his opponent begins to scream. Under such circumstances, one must stop the fight, or else the loser may be severely wounded or even killed.

The issue of how to stop a dog fight is not easy. Don't simply step in and try to separate the dogs. Doing this places you at risk from both rivals. Queen Elizabeth II of England learned this to her regret when she tried to separate two of her corgis who were fighting. In the end, she required several stitches in her hand. The royal dog trainer later provided a suggestion as to what she should have done. He suggested she should have grabbed one of the nearby silver trays and dropped it loudly onto the floor. Such a distraction will often stop the fight long enough to get the dogs under control. A bucket of water poured over the combatants or a spray with a hose will also work. I have found that a blanket or coat thrown over each of the fighters (not both under the same blanket), or sometimes even just thrown over the aggressor, will stop the fight and reduce the likelihood that you will get injured, though the coat or blanket may take some damage. Yelling or shouting won't help, since that simply suggests that you are barking and growling, and probably about to join the fight on one side or the other.

OTHER VOCALIZATIONS

Dogs make other sounds. Some are not specifically designed as words or signals. Nonetheless, you can often read these unintentional signs and gather some idea of what the dog is thinking. The most obvious of these is panting.

Panting: The characteristic sound of a dog panting, with its mouth open, tongue lolling out, results from a basic physiological need. It is simply the result of an attempt to control its body temperature. Evaporation of the moisture on the tongue and in the mouth cools the dog. Humans accomplish the same thing by sweating. The evaporating moisture from our skins cools us. Dogs, however, do not sweat from their skin the way we do, or horses do. The only place dogs sweat from is the pads of their feet, which is why an overheated or stressed dog may often leave wet footprints on the floor.

In human beings, stress, anxiety, or excitement can raise our body temperature. This is why people who are under pressure may begin to sweat. The same applies to dogs. So, when a dog who is not moving and is not exposed to warm conditions starts to pant vigorously, this means that he is excited due to stress (which may come about because of either positive or negative conditions). Although far from an intentional communication, we can read this as meaning, "I'm ready!", "Let's go!" or (especially if there are damp footprints on the floor), "I'm pretty worked up and tense about this."

Sighs: These vocalizations are simple expressions of emotion, which we can translate if we are alert to what is going on. Sighs are usually accompanied by the dog's lying down with its head on its forepaws. They can have two meanings, depending upon what else is happening and the presence of certain facial expressions. With eyes half closed, it is a sign of pleasure, meaning, "I'm content and am going to just settle down here." This may be seen when a dog has just eaten its fill, or often when a loved owner returns and the dog settles on the floor near him or her.

As in the case of some of the other vocalizations, the meaning can be quite different when there are other facial expressions or behaviors. If the dog settles into position with a sigh and its eyes fully open, this reverses the meaning. Now the sigh is a sign of disappointment that something anticipated hasn't materialized, best interpreted as, "I give up!" You might see this in a dog who has been hanging around a table where people are eating, in hopes it might be able to get a bit of food. If people finish their meal and it is obvious that no food is forthcoming, you might get the eyes-wide-open sighing pattern. If my dog Odin has been whimpering to get me to fetch his Frisbee off the shelf and take him out to play, but now he perceives I haven't responded and have settled down to my desk to work, he will often react with this characteristic disappointment sigh. My daughter Karen's dog, Bishop, has a more emphatic version of this sighing behavior. When he has been asked to settle down, or move, or to do something that he finds mildly annoying, he gives what could be called a sigh-snort. We interpret this as his version of "Oh, all right!"

7

Learning to Speak

Each species is provided with a "prewired" ability or predisposition to learn to understand and produce the language or communication behaviors of its own kind. Human language (as far as we know) is probably the most sophisticated and elaborate, and thus we probably have the most sophisticated form of this genetic predisposition. A child's development of language is an almost magical combination of this prewired ability working with the language in the child's environment. Consider the fact that the average high school graduate knows somewhere around 80,000 words. If we suppose that we start learning words at around one year of age, then that averages out to nearly 5,000 words learned each year—that's thirteen words learned each day. The most astonishing aspect of this language learning is not just its rate, but the fact that most of this language is not formally taught to the child. Obviously, this is why children who live in places where there are no schools can still proficiently speak their native tongue. Children simply imitate the vocal behaviors of whoever they are reared with. By ten months of age, when children are still babbling, they are already making sounds that allow a linguist to recognize the language spoken in their home environment. In other words, children living where English is spoken babble in English, while those living where Chinese is spoken babble in Chinese.

A fascinating example of how children simply absorb the language or languagelike behaviors in their environment comes from an event which began in October 1920, when a Christian missionary, Reverend J. A. L. Singh, was out on one of his usual soul-saving expeditions in the Bengali region of India. He would gather pagan volunteers from nearby villages, who were usually willing to listen to him preach as long as they were permitted to hunt between sermons. At the village of Godamuri, he was told a strange story about a *manushbhaga,* or man-ghost, that had been seen several times over the previous few years. It was usually seen in the company of wolves, who were going in and out of a giant, dead termite mound they seemed to use as a den. Reverend Singh had a hunter's blind constructed near the anthill, and shortly after dark he observed a wolf coming out of the mound. Singh calls it a wolf, but in his diary[1] he notes that it was probably a type of jackal, which is common to that region, rather than a true wolf. In any event, the "wolf" was followed by some others, and then came a grotesque-looking animal. It had the body of a human but walked on all fours, with the palms of the hands flat on the ground. The head seemed to be "a big ball of something covering the shoulders and upper portion of the bust." There was obviously a humanlike face visible under this ball. (The ball later turned out to be an accumulation of matted hair.) Following this animal came another one, just like it, only smaller in size. When Singh proposed to dig up the mound, the local villagers refused. They were afraid that disturbing the "ghosts" might bring some sort of retribution or curse down on them and their village. In the end, the Reverend went to another village, which did not know of the story, and found some more willing workers.

On the morning of October 17, the mound was dug up. As soon as the digging began, two wolves ran out and escaped into the jungle. A third chose to defend the den. Reverend Singh later said that it saddened him to have to kill her, since it appeared to be a divine act that she had chosen to keep these two strange creatures alive (although it did seem likely that she had originally brought them to the den to serve as food for her cubs). Inside the den they found two wolf cubs, and huddled next to them the two strange creatures, which turned out to be human children. The older girl was about eight years old, and they named her Kamala, while the younger was about two, and they named her Amala. Amala would die within the year; Kamala would survive until she was about eighteen.

For our purposes, the most interesting aspects of this case have to do with the behavior of the children. In addition to walking on all fours, they had other wolfish behaviors. They sniffed everything that they were given, and ate and drank, like dogs, from a plate on the floor. They preferred raw meat, and would growl, snarl, or snap at anyone who came close while they were eating. If they were frightened, they would back away, snarling and showing their teeth. Once Kamala became comfortable with her surroundings, she would sometimes pick up a toy in her mouth and run away with it, much like dogs do when they are playing with each other. She seemed to be trying to induce a canine-style chasing game.

Reverend Singh initially reported that the girls were mute, but by this he meant that they spoke no human language. They did make sounds, such as the sort of growling we've already noted. They also had a high-pitched whimpering sound, much like that of frightened or lonely puppies. They would occasionally make yipping sounds when they were excited, again like a playful puppy. But perhaps the most striking sound they made was howling. It started with a hoarse low voice, which gradually changed to a long loud wailing that had many similarities to the nighttime howls of wolves, jackals, and dogs. In the early days after their rescue, the girls would prowl around at night. During their prowling, they would stop to howl at regular times, usually around 10:00 P.M., 1:00 A.M., and again 3:00 A.M. Their vocal behaviors were exactly what we would expect if the children had been exposed only to the vocal sounds of wolves. It appears that in the same way that normally reared children imitate the sounds of the language spoken in their household, Kamala and Amala had learned to reproduce the sounds "spoken" in their canine household.

Whereas human beings instinctively learn language by duplicating the speech sounds in their environment, most animals do not have the genetically prewired predisposition to imitate vocalizations. Even if they had the physical capability to create human language sounds, they lack the instinct spontaneously to copy the words they hear spoken. This alone would make it unlikely that they would learn spoken language in the same manner that humans do. However, few animals have ever been treated exactly the same way that human children are. This brings us to the case of Gua, a female chimpanzee, who at age seven and a half months was separated from her mother in the spring of 1931 and given to Professor Winthrop Kellogg and his wife, Louise. The Kelloggs wanted to conduct an experiment to see if a chimpanzee reared like a human child, in a normal family setting, would develop human skills, including language.[2] This was not a totally bizarre expectation, since the DNA of chimpanzees and humans differs by less than 2 percent. Given this genetic similarity, it seems reasonable to suggest that raising a chimpanzee in a normal human environment and treating it as one might treat a human child might give it many more humanlike attributes and abilities, and this might even include some humanlike linguistic ability.

Gua was treated as if she were the younger sister of the Kellogg's nine-and-a-half-month-old son, Donald. Just like Donald, she was diapered, bathed, and powdered. When she ate, she was seated in a high chair and fed with a spoon. She was spoken to in the same way that Donald was, and for the nine months that she lived with the Kelloggs, they treated Gua just like a human child.

Compared to Donald, Gua developed much more quickly in terms of her motor skills. She could manipulate things earlier and better than Donald, and learned to walk and run sooner. The one area where Gua

began to fall behind rapidly was in her language skills. No special attempts were made to teach Gua language. It was expected that if she was
capable of learning language, she would learn it in much the same way
that human children do by imitating the language they are immersed in.

Gua did develop a means of communicating by gestures and body
movements. For example, when she saw a glass of orange juice on the
table, she went over and made lip-smacking or kissing motions on the
table's edge to show she wanted it. She would also point to things she desired, or simply to draw the attention of others to things that she found
interesting.

Gua did make sounds. They were not, however, word sounds, but
rather the typical vocal noises made by wild chimpanzees. The number
of these sounds never changed during the time that she lived with the
Kelloggs. In most cases it was quite clear what was meant by them, such
as a cry of pain when she hurt herself; a panic cry; a set of anger sounds;
excitement screams; and contentment grunts. Perhaps the most interesting feature of these sounds is the way in which two of them came to have
a new, extended meaning. The first is the food bark, which chimpanzees
give to alert members of their band that they have found something edible. The second was the "oo-oo" sound that normally means trouble or
expresses a moderate level of fear or anxiety. The food bark came to
mean the equivalent of "Yes" and would be given to questions like "Do
you want an apple?" or, "Do you want to go out?" The "oo-oo" sound
came mean "No," and might be given to questions like "Do you want to
have a bath?" or when asked if she wanted to go to bed for a nap with
"Do you want to go bye-bye?" She didn't learn human words, but rather
adapted her own chimpanzee language sounds to respond to and communicate with the humans around her.

Despite the fact that Gua did not make English-language sounds, she
clearly comprehended them, and could understand and correctly react to
over seventy words or phrases by the end of the nine months. It appeared that she was responding to key words in phrases, rather than the
whole sentence. For example, she would give the same food bark to the
question, "Do you want an orange?" and to the one-word query, "Orange?" The intonation—that is, the rising pitch of our voices—is what
allows us to know that a spoken word is a question. For Gua, intonation
was not so important, since the word "orange" spoken in a monotone
or even with a falling pitch at the end would produce the same response.

Gua sometimes did have problems in understanding, especially when
words were put together to make combinations that she had never heard
before. In one instance, after being taught the meaning of "Kiss Mama,"

she was then told to "Kiss Donald." It was obvious that the human child understood this request, since Donald promptly turned his cheek up to receive the kiss. Gua, however, acted confused by the new sentence.

Obviously, the Kelloggs' experience with Gua showed that animals do not usually spontaneously learn to imitate human language sounds. There was, however, an amusing demonstration of the fact that humans spontaneously imitate sounds in the world around them. Although Gua did not imitate English sounds, Donald rapidly learned Gua's full repertoire of cries, barks, and screams. Furthermore, he seemed to use them correctly, at least as defined by Gua's behaviors. So, humans can spontaneously learn to howl like a wolf or bark like a chimpanzee, but these animals do not seem to respond with similar imitation of our language sounds.

We have been discussing the limited range of cross-species "language" imitation by animals other than humans. However, there are rare examples where animals do learn to imitate the intonations of human speech so that they can vocalize a sort of "echo" of some human words. The "speaking" ability of parrots is an example of this. The one case of spontaneous verbal imitation of a word by a dog that I actually got to observe involved Brandy, a standard poodle owned by psychologist Janet Werker of the University of British Columbia. Brandy stayed at home during the day. Each night when the family members returned home, they would usually enter and greet the waiting dog with the word "Hello." Their salutation to their dog was always given in a cheerful, singsong tone of voice. After a while, the dog learned to imitate that two-syllable, singsong "Hello" with its own two-syllable version, "Arl-row." This became the sound which Brandy then gave spontaneously when greeting family members who were entering the house. This vocalization, however, was reserved specifically for family and was never given to strangers. Brandy appears to have added an English word sound to his existing Doggish vocabulary.

Though dogs cannot, or do not, usually imitate human language, most canines can learn to imitate the sounds of other canines. An interesting case of wild canines copying the sounds made by domestic canines comes from the early 1970s in the Canadian Yukon Territory. A wolf behavioral and biological project was underway there, and it involved the tagging and examination of a number of wolves. One pack—a group of four adults and two juveniles—was isolated and drugged so that they could be tagged for later identification, and also so that they could be examined by a veterinarian to determine certain aspects of their health. The vet was quite distressed when he examined the animals, since he found that three of the adults were showing signs of a respiratory disease. His concern was that this disease might be contagious, and returning the infected wolves to the wild might spread it to other wolves in the area, perhaps even to other species. The disease was potentially debilitating; if the three infected adults eventually died because of complications, he doubted that the remaining two juveniles and one adult would be able to survive, even if they escaped infection. Treatment, on the other hand, was simple. If the wolves could be confined for several weeks, a course of antibiotics would cure the disease. With this in mind, the six unconscious wolves were brought to a nearby settlement and placed in a fenced area that was part of a kennel where sled dogs were kept.

As we have already seen, wolves do not normally bark, except when they are young puppies and very excited. The dogs in the adjacent kennel area acted the way that dogs do, barking warnings and greetings. This transplanted pack of wolves then found themselves living in an environment filled with the sounds of domestic dogs barking. Over the course of the month or so that the wolves were kept in, their behavior began to change. In the last week of their stay, the researchers noticed that when people approached their enclosed compound, the two juveniles and one of the adult wolves would rush forward and begin to bark. The sound was a bit huskier than one normally hears in dogs, but the pattern of barking followed the usual alerting pattern of the domestic dogs, with bursts of around three or so barks, followed by a short silence, and then another burst of barks. It certainly seemed to the people present that these wolves were imitating the vocal sounds of the other canines in their environment.

Domestic dogs are at some disadvantage in the development of communication skills. If they have been allowed to stay with their mother and littermates until they are at least eight weeks of age, they should have learned the meaning of some basic canine signals, in the form of both vocal and body language. It seems clear that a great deal of dog communication is genetically prewired, but it also seems that other dogs modeling communications are needed to develop a full range of canine signals. Once the dog is removed from an environment where it is interacting with other canines and placed in a human family environment, it is "on its own" in trying to learn further communication skills. Dogs obviously won't imitate human language; but if they do encounter other dogs, they will often take that opportunity to learn to reproduce any additional canine vocalizations they might hear.

I have seen dogs imitating the barking behaviors of other dogs in various situations. For example, Karen and Joseph Moss owned a Gordon setter. This breed is not particularly noisy in most cases, and Sheila was a particularly quiet example of her kind. The Mosses' eldest daughter had been living on her own for a few years, but now had the opportunity to go for some advanced training at a school in another city. Since she would be living in a dormitory for the year, she couldn't bring her Airedale terrier, Argus, with her. Karen and Joseph agreed temporarily to "adopt" the dog until their daughter returned. Argus was a typical terrier: He barked when people came to the door, he barked when people left the house, and he barked just for the joy of hearing himself bark. As the weeks passed, Sheila's behavior began to change. Now, when Argus barked at the door, so would she, and sometimes, when Argus was simply charging around the house barking at phantoms and fantasies as a game, Sheila would join in with her own sonorous bark. Long after Argus had left to rejoin his mistress, Sheila still continued the barking behavior she had learned from her terrier friend.

My dog Odin has a "request bark," which he uses to ask to be let back into the house after being out in the backyard for a while. It is a single bark, followed by a long pause of anywhere between thirty seconds to two minutes before it is repeated. I rewarded this bark when he was six or seven months old by quickly responding and letting him back into the house. It is not only a bark that has been learned for use in this way, but it actually sounds different from his other barks—somehow more artificial or forced, since it ends with a sort of raspy finish. When Dancer arrived at my house as an eight-week-old puppy, I would let him out with Odin to eliminate, as part of his housebreaking. It took less than a week for Dancer to develop that same bark. The time pattern is the same, but being only a puppy, the pitch of his bark is much higher. Still, even with the tonal differences, it sounds forced and artificial in the same way that Odin's bark does, with that same raspy finish to each single utterance. Now, at the age of twelve weeks, Dancer will use this bark even if he's alone outside. He has learned to imitate the sound that Odin makes, and has also learned the communication value of this sound. The adult dog had taught the puppy a word or a phrase in his dialect of dog language.

Sometimes dogs in the same household will imitate each other to the degree that they develop a sort of common Doggish dialect. An example of this involves excitement noises. Remember that there are three different sounds that characterize excitement in dogs. These are the excitement whine, the moan-yodel ("Vowel-wowel-owel-wowel"), and the howl-yawn (a breathy "Hooooooo-ah-hooooo"). All of these sounds are made as the dog stares directly at you and twirls around to show its happy excitement. Each dog seems to decide on which of these sounds it makes in excited anticipation and it does not seem to depend upon their breed. The odd thing, however, is that dogs that live together seem to imitate each other so that all are using the same sound. Thus, I have an acquaintance who owns four flatcoat retrievers, and all of her dogs make the howl-yawn sound. I know another person with three dogs of different breeds—a Pekingese, an English springer spaniel, and a flatcoat retriever—and all (including the flatcoat) make the moan-yodel sound. I made an informal survey of sixteen people who own more than one dog, and whose dogs have all grown up from puppyhood in the same house. Twelve of these people reported that all of their dogs use the same excitement sounds, regardless of breed. This strongly suggests the dogs are copying each other's sound patterns.

Since dogs only seem to copy the sounds of other dogs, this can make it fairly difficult for a human who wants to train his or her dog to make sounds. Of course, as an amusing trick, many people teach their dogs how to "speak," which really means to bark on command. But these voluntary barks sound qualitatively different from a spontaneous bark. They seem somehow emotionless, and sometimes as if not fully voiced. The same can be said for the bark that police and protection dogs learn to give to announce the location of where someone is hiding. One master of a police dog told me that his dog's detection bark sounds "faked" to him. "It's not like a real bark, which has some passion; but most people, who aren't used to listening to dogs, can't tell the difference. Maybe they're just too worked up over a dog barking at them to tell the difference."

Some dogs can be taught to make specific sounds under certain conditions. These sounds can range from simple barks, moans, or play-growls through more complex sounds, which may sound like yodels or attempts at speech. Training a dog to make these sounds is not done by making barking sounds with the expectation that the dog will copy them. Instead, when the dog makes the sound that you want to put under control, you give the command and reward the dog. The problem, obviously, is that you can't reward the dog for making a particular sound unless it first makes that sound spontaneously. Take the case of Ann, a woman who lived alone in the city and wanted to teach her extremely friendly chocolate Labrador retriever, Caesar, to bark on signal. The problem was that Caesar joyously greeted everyone who came near and never barked at strangers. Ann felt a bit insecure and defenseless in her neighborhood. She felt that if Caesar could learn to bark on command, the sound could be used to ward off strangers who might approach her on the street or come to her door. She believed that having a "protective" dog who barked would make her feel much safer.

When Ann came to me, we first had to decide on the commands that we would use. Someone who hears you tell your dog to "Speak" and then hears barking is not apt to be very worried by the performance. Instead, we chose the word "Protect!" as the command to start the barking. We felt that if a menacing person approached and the dog started to bark after the command "Protect!", they would be more likely to think Ann had an attack-trained dog and would back off. The command to stop barking was to be "Keep watch!" which would sound as though the dog, although not barking now, was still in a guarding mode. Of course, to Caesar, "Protect!" would mean nothing more than "Speak" might to another dog, and "Keep watch!" would mean the same thing as "Quiet."

Deciding on the commands turned out to be a lot easier than actually getting Caesar to speak. The first step toward getting a dog to bark on command is to find some situation where the dog will spontaneously do so. Once he is actually barking, you can give the command and praise the dog. Usually this is accomplished easily by placing the dog on leash and having someone knock on the door or ring the doorbell. As soon as the sound is heard, the dog's master is supposed to call out excitedly, "Protect!" If the dog barks, he gets a treat. Unfortunately, no matter who came to the door, Caesar wouldn't bark.

Next, we increased the implied threat to try to get him to bark. We repeated the person coming to the door setup, only this time Ann was instructed to act alarmed. She fluttered around in an agitated manner, waving her arms and calling out, "Caesar, protect!" If he barked, he was to be praised and would get a treat. Again, Caesar didn't bark.

Then we moved to a strategy that occasionally works. Although dogs don't imitate human speech, they do imitate other canine sounds, as we have seen. Sometimes it is possible for a human to get a dog to bark by the human reproducing the sound of dogs barking. The best way to do this in my experience involves making panting or huffing sounds, followed by a woofing sound that mimics a muted bark. The dogs don't try to imitate this human bark, but often get excited and begin spontaneously to give their own version of a "call the pack" alarm bark. Thus, Ann was to go: "Caesar, protect!—huff, huff, huff, woof, woof, woof!" when the person approached the door. Caesar did seem to get quite excited by all of this, but he still didn't bark.

In the end, we used some more provocative measures in a sort of game format. We set up a situation where Ann was outside with Caesar on leash. I then approached him with a dust mop. I jumped around, shouting, and waved the mop in front of Caesar's face while Ann called: "Caesar, protect!" This produced Caesar's first bark of the day, but it was obviously a bark of fright, since he was straining backwards on the leash. After he barked, Ann immediately praised him: "Good dog! Good protect!" At that moment I quickly backed up to give Caesar the confidence to respond the next time. After a brief pause, I approached again, shaking the mop toward Caesar's face while Ann instructed him to "Protect!" He again tried to retreat and now gave two quick barks. While Ann was praising him, I backed off again.

After two more tries, it seemed that Caesar got the idea that his bark made "that thing" go away. Now as I retreated, he decided that he wanted to "get that thing," and began straining forward on the leash as I retreated. As he pulled forward, he was barking loudly at the mop. Labs are bright dogs. Now that he was making agitated barking sounds and being praised for it, he reasoned out that it was his vocalizations which made that thing retreat and made Ann happy. Once we had him responding reliably to "Protect!" we moved on to the next phase. After Caesar had been given a chance to give a series of very excited barks on cue, Ann then placed her hand over his muzzle and quietly said: "Keep watch!" Now when he stopped barking, he was praised.

A day or two later, we repeated the practice sessions, only now I approached Caesar with an umbrella. When I opened it in his direction, Ann said: "Protect!" Now when Caesar barked, I closed the umbrella immediately and backed off to give the dog more confidence. When I was at a distance, she said, "Keep watch," and when he became quiet, she gave him a treat. Later on, we had a different person approach with their coat held over their head. By now, Caesar had learned that "Protect!" meant "bark." From that point, it was a simple matter to transfer the training to people approaching without dust mops, umbrellas, or waving coats. By the end of a week, Caesar was barking on command when someone arrived at Ann's door. To a stranger, Caesar's bark and his straining at the leash are quite convincing. The city often feels like an unsafe place for a woman living alone, and Caesar's barking gives Ann the extra margin of security she felt she needed. Of course, when faced with Caesar's frantic barking, what the potentially dangerous person doesn't know is that a big smile and praise in the form of "Oh what a good protect—what a good bark, Caesar!" would change the noisy threatening behavior into a happy approach with wagging tail!

Even though dogs do not have the ability or an instinctive predisposition to imitate human sounds, we have seen that they have a useful and meaningful vocabulary of vocal sounds, some of which may have been learned from other dogs. Sounds, however, are not the only medium that dogs can use to "talk." There are many other ways of sending messages which do not involve sounds, but which serve as important channels for canine communication.

8

Face Talk

uman beings communicate a great deal with their faces. The language of the face speaks of a broad range of emotions and even of very subtle intentions. Indeed, "face talk" is so informative that some people such as card players, negotiators, news reporters, and certain business people often have to school themselves *not* to say too much with their expressions.

Humans can tell lies with their faces. Although some are specifically trained to detect lies from facial expressions (e.g., Secret Service agents, specialized police officers, and clinical psychologists), the average person is often misled into believing false and manipulated facial emotions. One reason for this is that we are "two-faced." By this I mean that we have two different neural systems that manipulate our facial muscles. One neural system is under voluntary control and the other works under involuntary control. There are reported cases of individuals who have damaged the neural system that controls voluntary expressions. They still have facial expressions, but are incapable of producing deceitful ones. The emotion that you see is the emotion they are feeling, since they have lost the needed voluntary control to produce false facial expressions. There are also clinical cases that show the flip side of this coin. These people have injured the system that controls their involuntary expressions, so that the only changes in their demeanor you will see are actually willed expressions.

The reason that people get away with lying is because the involuntary neural system has more control over the upper part of the face, while the voluntary system has more control over the lower (probably because activities like eating and speaking are voluntary and require conscious control over the mouth region). This is important because human beings apparently pay more attention to the signals from the lower half of the face when trying to interpret another person's feelings. People who are trained to look for deceptions, and people who are just naturally good at this skill, read the whole face, including the eyes. It is possible to tell a false smile (which can represent conscious deceit or simply lack of true emotion) by the fact that it exclusively uses the muscles in the lower face, which only affect the shape of the mouth. A true smile also involves the muscles that are higher up and serve as "cheek-pullers." In the true smile, the action of this muscle puffs up the cheeks a bit and narrows the eyes, while the false smile involves only turning up the corners of the eyes.

Some involuntary muscles around the mouth do produce expressions that are difficult to fake. For example, grief and sorrow pull the lip corners down, without moving the chin muscles. Scientific studies show that fewer than 10 percent of people can do this voluntarily. Another mouth sign that is involuntary is the tightening of the lips, which is a dependable sign of anger. The lips don't pull in or press together in a noticeable way, but actually appear to become thinner, almost as if a bit of the fleshy part of the lip has rolled into the mouth. This gives a very subtle and involuntarily stern look to mouth.

To camouflage their fear, anger, or guilt, accomplished liars who know that these automatic mouth movements cannot be hidden attempt to cover these signals with another strong emotion. Using forced laughter is a common trick; another is indignation. The Israeli interrogators who questioned the war criminal Adolf Eichmann reported that he used this technique. One later wrote that whenever Eichmann burst out with an indignant "Never! Never, Herr Hauptmann!" or "At no time! At no time!" he knew that he was lying and was trying to obscure any facial signals that might betray this.[1]

The facial expressions of dogs, especially those involving the lower part of the face and the regions around the mouth, are similar in many ways to those of humans, but they are more limited in their range. Dogs either don't have, or don't use, the voluntary neural control system to shape the expressions conveyed by their mouths. This doesn't mean that dogs are incapable of telling lies; only that dogs don't use their mouths and their facial expressions to do so. Another limitation on the facial expressions of dogs is that their muzzle is constructed differently from the human mouth. It was designed for a more limited set of uses and this restricts the number of possible expressions.

Almost all vertebrate animals, except man, have a muzzle. Lions, bears, birds, alligators, and dogs all have a projecting mouth, which forms their muzzle. This is a piece of basic survival equipment. It lets these animals grab, gnaw, or nip at others, and in most other animals the mouth is also used to actively seize food. A cow will munch grass, while a tiger will bite and tear at its kill. Humans do not need a muzzle for this purpose because we, and most other primates, use our hands to bring the food directly to our mouths.

The muzzle serves as a powerful weapon. An elongated muzzle makes room for lots of teeth. Teeth are, of course, the basic tools used by carnivores, such as dogs. The muzzle positions those teeth so that they extend outward. This allows the mouth to close like a trap. Powerful muscles are attached to the hinge at the rear to exert great force when biting. An average dog will have a bite strength that exceeds 900 pounds per square inch. Even toy dogs have bite strengths of 700 pounds per square inch or more, while Labrador retrievers, who can carry a bird so gently that it is delivered to their masters without its feathers being ruffled or its skin broken, can still bite with a pressure of over 1,000 pounds per square inch. Some of the bigger dogs with square faces, like the mastiff or the Rottweiler, have measured bite strengths in the range of 2,000 pounds per square inch, which would make them formidable opponents when attacking.

Although it is important to have muscles to make the bite strong, the musculature of the lips is not nearly as important, since the lips do not play a big role in killing, gnawing, or eating food. Canines drink by lapping with their tongues, so they don't need lip control to allow sucking or conforming the lips to the shape of container that humans have. They do need to suckle as young puppies (until around six weeks of age); however, the muzzle of a puppy is much shorter and smaller, which enables them to form the narrow mouth opening needed to create a vacuum for sucking. Limited lip control restricts the variety of facial expressions that dogs can make with their mouths, but they still retain enough flexibility for a range of communication signals. In fact, the mouth is probably the dog's single most important means of expression—and I don't mean to restrict this simply to sound.

MOUTH SHAPE

The language of mouth shape in dogs follows the general pattern of other gesture-based communication systems, in that signals are reserved for a few major issues. The mouth gives information about anger, dominance, aggression, fear, attention, interest, or relaxation. Let's look at some of these expressions.

Mouth relaxed and slightly open; tongue perhaps slightly visible or even slightly draped over the lower teeth: The sign of a content and relaxed dog. It is the dog equivalent of the human smile, and has been recognized for centuries by humans. In ancient Egypt, traditional children's toys were made in the shape of animal faces, and the most common form was the "smiling animal." Such toys featured an extended tongue, lapping out over the front of the partly open mouth, thus mimicking the smile of a dog. The translation of this expression for dogs is "I'm happy and relaxed," "All is well," or "I see no threats or problems around me."

Mouth closed, no teeth or tongue visible: The simple act of closing the mouth changes the meaning of the dog's expression. The closed mouth expression is usually associated with the dog looking in a particular direction, and the ears and head may lean slightly forward. This is a sign of attention or interest. The smile is now gone, mostly because the dog is appraising the situation, trying to determine the meaning of what he is observing, and perhaps evaluating which actions he might take. The dog is no longer passive, but also not worried or annoyed. Thus, this expression takes on the meaning of "This is interesting" or "I wonder what's going on over there?"

Warning signals involve curling or pulling the lips to expose the teeth and perhaps the gums. The rule of thumb for mouth expressions made by dogs is simple: the more the teeth and gums are visible, the more likely the dog is signaling aggression. This very functional signal has evolved because it shows the dog's weapons (teeth) and lets the viewer understand there will be negative consequences if the warning is not taken seriously. Thus, the other party has a chance to back down and leave or to make some pacifying gesture. If fights can be avoided by such signals, then survival, not only of the individual, but of the pack and the species will be more likely. Actual physical fights cause injuries, and an injured animal may die, leaving the pack weaker and perhaps puppies untended. Even if it survives the injuries, until healing has been completed, the animal will not be as effective a hunter, defender of the pack, or provider for the young.

Lips curled to expose some teeth, mouth still mostly closed: The first sign of annoyance or threat in a dog. The dog is not worried, and may be silent as it looks toward the source of annoyance or alternatively makes a low, rumbling growl. This is basically a signal to a nearby dog that greater distance and less social interaction would be appreciated. It is not just a simple request but clearly the first sign of menace or threat. In everyday human language, it translates into "Go away! You're bothering me!" or "Back off! You annoy me!"

Lips curled up to show major teeth, some wrinkling of the area above the nose, mouth partly open: We can translate this signal as meaning, "If you do something that I might interpret as a threat, I will bite." Note that this expression only tells you about the intentions and feelings of the individual who displays it. It says nothing about the underlying causes for giving the threat signal. The threat might be an expression of social dominance given by a confident and high-ranking dog, but, as we'll see below, it may also indicate fearfulness. In either case, pressing a dog by drawing too close in this situation may lead to an aggressive attack. Ceasing to approach, standing still, or backing away makes much more sense.

Lips curled up to expose not only all the teeth but also the gums above the front teeth, with visible wrinkles above the nose: This is the last warning that a physical attack is not only possible but might be triggered momentarily: "Back off or else!" The full threat display indicates that the dog is ready and willing to deliver a violent attack.

If you are ever confronted by this display, even if you are frightened, you should not turn and run. All dogs have a genetically programmed pursuit response which makes them instinctively chase and perhaps bite things that are running away from them. In this case, even if the threatened aggression is based upon the dog's fear, rather than angry confident dominance, the level of arousal is so high that running or even moving quickly can provoke pursuit and attack responses. Later on, I'll tell you how to communicate that you are no threat to dogs who are giving this extreme warning.

FEAR, ANGER, OR DOMINANCE?

In all of these expressions, I have only concentrated on the degree of the threat that the dog was communicating. The *trigger,* or reason, for these warning signals is another matter. Threats can be triggered by an attempt to assert social dominance; they can be triggered by anger or annoyance; and they also can be initiated by fear. Knowing the nature of the emotion that the dog is signaling is important because it predicts how the dog will behave. Fearful dogs act in a manner different from dominant dogs who feel in control of the situation. In response to a challenge, an angry or annoyed dog who is confident about its status will only continue threatening until the individual bothering it actually goes away. Then the dog will happily return to quiet, sociable behaviors. A fearful dog, however, will often remain fearful for a long time. The incident will disturb its confidence and it can easily be jarred into responding aggressively again by anything unexpected happening nearby. It may also turn and run away in panic the moment the impending confrontation has ended. Therefore, it is important to understand the motivation behind the aggressive message.

While the level of aggression is signaled by the degree to which the teeth and gums are exposed, it is the nature of the lip curl and the shape of the mouth opening that tells you whether it was anger and dominance or fear that triggered this expression. Take a look at Figure 8-1. The top face shows a dog that is signaling aggression but is still interpreting the situation and awaiting further events. If we move to the left-hand column, we veer toward depictions of dominance and anger motivating the aggression. The right-hand column represents fear and aggression. As you move down the columns, the intensity of the emotions (anger or fear) increases and the likelihood of aggression also increases.

Looking carefully at the shape of the mouths, those associated with anger and dominance are quite different from those signaling fear. Anger produces an open mouth in which the contour is sort of C-shaped, and the teeth that are most visible are the big front canines. The back teeth in an angry dog are barely seen. The shape of the mouth for fearful dogs is elongated, as though the inside corner of the mouth has been pulled backwards, stretching out the mouth opening. Because of this elongation, the rear teeth become more visible.

The mouth signals in this figure are amplified by ear and eye signals that we will talk about later. For the moment, simply notice that the dominant dog's ears are forward and perhaps slightly spread to the sides, while the more fearful and submissive dog's ears are slicked backwards and flattened against the head. The eyes on the dominant dog are also larger, with a more intense stare, while the eyes on the more submissive and fearful dog are more slitted and less wide open.

Thus, the faces in the left column show aggressive anger. They are saying (with increasing intensity): "If you continue to bother or challenge me, I will hurt you." The faces in the right column are more fearful; they say: "You frighten me, but I'll fight if I'm forced to." Dogs showing fear-based aggression are not less likely to bite. A fearful dog will defend itself perhaps with even more intensity than a dominant dog, since the frightened dog is worried about its safety and survival. Regardless of whether the aggression was stimulated by fear or anger, the more intense the emotion (the farther down the column), the more likely the dog will carry out its threat.

Head positions: There is another dimension to canine mouth signals. This is simply whether the mouth is turned toward you or not. For canines, their only really dangerous weapon is their teeth. A dog that looks directly at another individual points its weapon at them. This may produce the same effect as a human produces by pointing a gun at someone, arousing fear and defensiveness in the target individual. A dominant or threatening animal will use the gesture of pointing their muzzle at some-

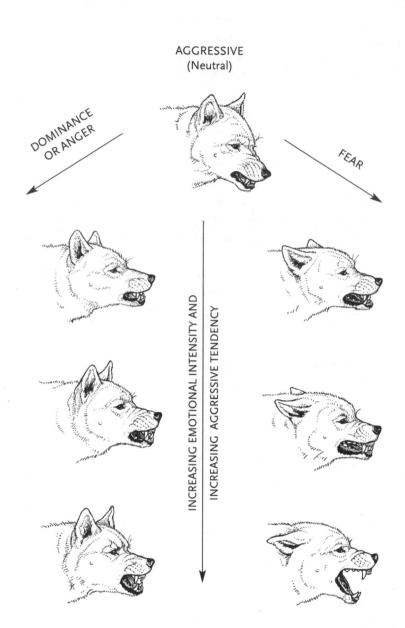

AGGRESSIVE
(Neutral)

DOMINANCE
OR ANGER

FEAR

INCREASING EMOTIONAL INTENSITY AND
INCREASING AGGRESSIVE TENDENCY

Figure 8-1 The top face represents a wary aggressive signal. Moving down the left column represents increasingly aggressive signals motivated by dominance, while moving down the right column represents increasingly aggressive signals motivated by fear.

one as a threat. Alternately, a dominant animal may calm the fear of a more submissive animal by slightly turning its head to the side, pointing the mouth away to show that it has no intention of attacking.

When a less dominant animal approaches one that is higher-ranking, the subordinate animal may approach with his head down and only an occasional quick pointing of his muzzle in the direction of the more powerful animal. It may respond to a direct stare from the dominant individual by turning the head so that his mouth is now directed to the side, and away from the other dog. In effect, turning the mouth away is equivalent to saying, "I've holstered my weapon and am not pointing it at you. You can rest easier now, there will be no fight."

Dog yawns: This is probably one of the dog signals that humans are most likely to understand. When a dog is seen yawning, most people think the dog must be tired or bored and dismiss it as a minor or unimportant signal. This is not the case.

Physiologically, a dog's yawn has the same characteristics that a human's yawn does. It supplies additional oxygen to the brain, which can help keep you awake or awaken you more quickly after you've been asleep. Because of this, it is not surprising that dogs, like people, will yawn when they are tired. However, for dogs, yawning has a whole series of other meanings. A dog that is under stress will typically yawn. In dog obedience classes, I have often seen a dog yawn immediately after its master scolded it for something or gave it a very harsh correction. When training their dog to sit and stay or lie down and stay, people are often unsure whether the animal will remain in place or will break and run toward them (or toward the other dogs) the moment they've moved a few feet away. For this reason, novice dog trainers often use a very harsh, threatening tone, instructing their dogs to "Staaaaaaaay!" using their best imitation of the Voice of Doom. This tone of voice implicitly suggests that the dog just might die if he moves from his place. In a beginners' class, you will see many of the dogs left in a sit-stay position yawning, while their masters stand across the room from them. When the owner is taught to use a more friendly tone for commands, the yawning behavior usually disappears. In this sense, yawning might best be interpreted as "I'm tense, anxious, or edgy right now."

One of the most interesting uses of yawning is to send a pacifying message. The Latin root of the word "pacifying" gives us a good indication of its meaning, since it comes from *pax,* meaning "peace," and *facere,* meaning "to make." The yawn contains no elements of fear, dominance, or aggression. It is the exact opposite of a threat. When a dog is being menaced by another dog's aggressive signals, the target of this display may simply respond with a yawn. While the dog's human companion may view this as a sign of nonchalance or bored confidence, it is in reality sending a pacifying message. At the same time, yawning is not a sign of submission. The threatening dog will often break off its aggressive display immediately upon seeing its target yawn. It may even act a bit uncertain or unsure about what to do next, and then begin to initiate tentative greeting and approach behaviors.

There is another situation where yawning can occur, but in this case it is a pacifying gesture given by a dominant dog. Suppose that a dominant animal is approaching a submissive dog who may be fearfully protecting something like food. In these circumstances, the dominant dog may yawn, perhaps as a sign of friendly unconcern. This does seem to have a calming effect on the fearful dog.

Dogs do read yawning signals in humans. I was once a guest on a television show where we were to discuss the personalities of people who pick specific breeds of dogs. Several people had been asked to bring their dogs so they could be interviewed about their relationships with their pets and why they chose them. When I was ushered onto the set, three sets of dogs and owners were already seated there. As is typical in such situations, the dogs were a bit fidgety with all the lights, movement, and strange people around them. My place on the set was supposed to be next to the host of the show. On my other side was a woman with a large Rottweiler. As I sat down, the Rottweiler gave a low, throaty growl, began to curl his lip, show his teeth, and stare directly at me. Apparently, he was already uncomfortable in this strange situation, and having a stranger blithely sit down so close to him was the last straw. He was now indicating that he wanted me to back away and give him some space. Unfortunately, I couldn't do that. Even worse, there was only a minute or two before we were supposed to go on air, so there was not time to go through the usual greeting ritual that I use when meeting an unfamiliar dog. Since there were few other alternatives, I simply looked off to the side, breaking eye contact with the dog, and gave a large and exaggerated yawn. The dog looked at me and began to blink. I blinked back, and the dog settled quietly down to the floor with his head on my shoe. The threat had been defused.

Since that time, I have sometimes used yawning or suggested it to others faced with potentially aggressive situations. Generally speaking, a yawn, followed by some other non-threatening greeting response, has either caused the dog to whom it was addressed to cease hostilities or to tone down its aggressive display. Yawning in public may be viewed as a relatively meaningless (or impolite) behavior among humans; it is conversation and conciliation when used by or directed to dogs.

Licking: This is one mouth behavior that is even more misunderstood by the general public than yawning. Ask anybody what a dog means when it comes up and licks your hand. Mothers explain this behavior to their young children with words like "Look, dear, Lassie is giving you dog kisses." Unfortunately, this is often incorrect. Licking behaviors can mean many different things, depending upon the context. They can't be interpreted simply as affection. Each and every licking behavior must be interpreted in terms of the form that it is taking and the situation at the time.

Licking differs from kissing in a variety of ways. The actual form of the behavior is quite different. When kissing, humans and non-human primates, such as chimpanzees, apply their lips to each other. Dogs are not making lip contact; they are licking. Kissing is most often applied directly to the face and sometimes to the hands in normal social contact. While dogs will lick faces if they get the chance, they will also lick hands, feet, knees, or whatever they can get close enough to get their tongues on.

Kissing usually takes place as part of a greeting ritual in humans and primates. You kiss Aunt Sylvia or your spouse's sister not as an expression of romantic love, or even necessarily as an overt sign of affection, but rather as part of a greeting behavior. Kissing often has no more significance than a handshake in this situation. The same thing occurs in chimpanzees. Dogs that are familiar with each other may also lick each other's faces when greeting. However, dogs often go well beyond the face. When greeting each other, dogs begin to sniff any moist membranes, where odors are strongest—usually around the mouth, nose, anal regions, and urogenital areas. These greeting and identification sniffs may turn into licking as well.

Humans use kissing as part of their sexual behavior as well. It is when engaged in sexual activities that humans may avidly kiss parts of the body other than the face and the hands. Kissing seems to also be used by chimpanzees in this sexual manner, mouth to face, mouth to mouth, mouth to hands, mouth to genitals. For dogs, too, licking is part of the sexual ritual. Dogs often explore each other as part of their mating behaviors, even more vigorously than when they are greeting each other, licking each interesting area of their partner's body in the process.

While kissing in humans and licking in dogs do seem to have parallels in these aspects of communication, for canines, there is much more social significance to the lick. Licking can communicate information about dominance, intentions, and state of mind, and, like the yawn, is mainly a pacifying behavior. One thing in common among all pacifying behaviors is that they contain elements of puppylike behavior. Juvenile behavior is the canine equivalent of a "white flag." Most adults tend to nurture the young of their own species, and there appears to be a strong inhibition against actually attacking them. Thus, non-dominant, frightened, or weak adults will adopt juvenile postures and perform infantile actions to avoid aggression. These behaviors usually soften the mood of the threatening animal and will normally avert any sort of physical attack. Many aspects of pacifying behavior contain forms of licking. It makes sense to look at the early lives of dogs to interpret what these signals were meant to communicate in their earliest stages.

Dogs are first introduced to licking by their mother immediately after birth. Once the puppies are out of the birth canal, the mother licks them to clean them up. This same licking also serves to stimulate their breathing. In the days that follow, she continues to lick the puppies to groom them. Much of her licking is focused on the anogenital region to stimulate urination and defecation. I think that a convincing case can be made that the mother loves her puppies; however, when she is licking them, she is using her tongue to groom them, not to kiss them. Because she loves her baby, a human mother may wash her infant and change its diapers to keep it clean. Yet I am sure than no one would say changing a diaper is equivalent to kissing. Dogs effectively keep their puppies clean and remove their urine and feces with their tongues because they simply don't have the dexterity to do this with their paws, and certainly don't have the ability to diaper and bathe them.

As the puppies grow older, they begin to lick and clean themselves and their littermates. This mutual licking and grooming serves social functions. Obviously, it helps keep the puppies clean, but in the process it helps strengthen the bonds between the puppies. The actual mechanism that builds this affection is mutual satisfaction. A puppy can have companions get at those hard-to-reach places, like ears and backs and faces, and can pay them back by licking their littermates in their inaccessible regions. Since friends and familiars groom friends and familiars as a considerate gesture, the very act of licking another dog develops significance as a means of communication. Licking thus shifts from being a utilitarian and useful act to becoming a ritualized gesture. The meaning of this gesture at this time in a puppy's life involves goodwill and acceptance. In effect, each puppy is saying, "Look how friendly I am." As the puppy matures, the message sent by licking continues to be friendly but is widened to also mean, "I'm no threat," and perhaps the submissive plea, "Please accept me and be kind."

Licking takes on a further meaning a bit later in the puppy's life, usually around the time that it is becoming less dependent on its mother's milk. In the wild, when a mother wolf returns from hunting, she will have already fed herself on her quarry. When she enters the den, the puppies gather around her and begin to lick her face. To a romantic, this may look like a loving greeting with all of the puppies overjoyed at mother's return after her absence of several hours. They are seen as simply kissing her in happiness and relief. The actual purpose of all of this face licking, however, is much more functional. Wild canines have a well-developed regurgitation reflex, and the puppies lick their mother's face and lips to cause her to vomit up some food. It is most convenient for the mother to carry food in her stomach rather than trying to drag things back to the den in her mouth. Furthermore, this partially digested material makes ideal dining for young puppies.

It is interesting to note that our domestic dogs actually have a reduced sensitivity for their regurgitation reflex in comparison to wolves or jackals. Puppy-induced regurgitation is not as often seen in dogs unless the pups are not being fed well. When it does occur, it is more likely to occur in sharp-faced breeds that appear to be more similar to wild canines, such as the wolf.

Understanding the development of licking behavior helps to interpret another place where it occurs. Face licking in adult canines can be a sign of respect or deference to a more dominant dog. The dog doing the licking usually lowers its body to make itself smaller, and looks up, adding to the effect of juvenile behavior. The dog receiving the face licks shows its dominance by standing tall to accept the gesture, but does not lick the other dog in return.

Now when your dog tries to lick your face, you should have a better idea of what he's trying to communicate. He may simply be hungry and asking for a snack. Obviously, you won't regurgitate some food at that signal, but you might respond affectionately and perhaps give him a treat, such as a dog biscuit. He may be communicating submission and pacification—the adult version of goodwill in puppies. Basically, he is saying, "Look, I'm just like a puppy who is dependent on big adults like you. I need your acceptance and help." Alternatively, he may be showing respect and deference to you as a more dominant dog in his pack.

It is often a stressed and fearful dog who is exhibiting licking behavior, and these behaviors have become so ritualized that an anxious dog

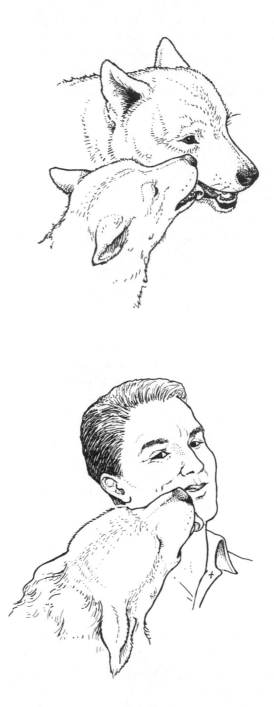

Figure 8-2   Licking behaviors are not "kissing," but can be pacifying and submissive signals, signs of respect, or simply a request for food.

may lick even when there is no dog or person close enough to be licked. It may lick its own lips, much the way stressed humans bite their lip. Sometimes, the dog will simply extend its tongue quickly and appear to be licking the air. At other times, the dog may drop down to the floor and nervously lick at its own paws or body.

I frequently see this kind of lip licking or air licking on the first day of a beginner's dog obedience class. The dogs are often stressed (perhaps because their handlers are a bit nervous), the surroundings are strange, and there are unfamiliar dogs in the room. As the classes proceed and the room, the situation, and the other dogs all become familiar, the licking behavior quickly disappears. I have been told by veterinarians that they often observe the same behavior in their consulting rooms. The dog licks the air and its lips while it seems to fret about this new environment, filled with strangers and unknown future events.

Licking, then, is a complex signal, and clearly not always the canine equivalent of kissing. It does send some important social messages, which we can read by reading the pattern and context of the licking behavior. However, since none of these messages is hostile in any way, I have no qualms in joining with everybody else on the planet by telling my grandchildren they are being kissed when my dogs lick them. It is certainly as harmless a myth as Santa Claus or the Easter Bunny, and it brings with it an equivalent dose of joy to the individual being licked.

9

Ear Talk

Although the dog's ability to communicate is reduced because its mouth is lacking in flexibility and control, it surpasses people in its ability to communicate with certain other parts of the body. In humans, for example, ears are not very expressive. Although I once had a childhood friend who delighted us all by wiggling his ears on request, most of us don't have much voluntary control over our ear shape or position. Our ears are effectively fixed into one position and have one fixed shape, which makes them useless for outgoing communications. Dog's ears, however, are well suited for sending messages.

PRICK-EARED DOGS

Dog's ears come in many different shapes. Some shapes are considerably better for communication purposes. Let's start by considering the most expressive ears in dogdom. All of the wild canines, and many domestic dogs, have pricked ears. These stand up and are visible from a fair distance. A moderate amount of mobility was built into the canine ear to allow it to rotate to catch significant sounds. Ear movement is more subtle than that of the whole head and thus less likely to give away an animal's hiding position. Evolution, being an opportunist, took advantage of both the mobility and the visibility of ears to create a channel of communication.

Although the positions of dogs' ears give very significant signals, they should be read in the context of the whole activity of the dog. Used in conjunction with other signals, they greatly add to the clarity of the message and permit the addition of certain nuances of meaning.

If you are confronted by a dog that is showing its teeth, wrinkling its nose, and growling, it is really important to look at ear positions if you want to understand its motivation. Usually, people focus on the fact that either they can see teeth bared or they can't. The subtle face talk signals associated with the curl of the lips or the shape of the dog's mouth are easy to miss when your adrenaline starts to rise because a snarling dog has made you uneasy. A shift in ear positions, however, is easy to see, and can modify our interpretation of the apparent threat. Let's start by looking at some non-aggressive ear positions.

Ears erect or slightly forward: An indication that the dog is studying the environment for information, or a sign of attention when alerted by a new sound or sight. It can best be interpreted as "What's that?"

The message sent by erect ears is a bit different when accompanied by a slightly tilted head and a relaxed or slightly open mouth. In this context, the message can be interpreted as "This is really interesting." This signal is often sent when the dog is observing a novel or unexpected event.

If we close the mouth and open the eyes a bit, the meaning changes subtly again. Now the message becomes "I don't understand that" or "What does this mean?" Under these circumstances, there may also be a slight slow waving of a tail held somewhat low. When this same ear signal is accompanied with bared teeth and wrinkled nose, it is an offensive threat by a confident dog. It really means, "I'm ready to fight you, so consider your next actions carefully."

Ears pulled back flat against the head: Associated with bared teeth, this is the sign of an anxious dog who is saying, "I'm frightened, but I will protect myself if I see you as a threat." This pattern of ear position and facial expression is usually found in a less dominant dog who has been presented with some sort of challenge that has it worried.

If the ears are held flat, the mouth is drawn back, the teeth are not visible, and the forehead is smooth and free from wrinkles, then this is a pacifying or submissive signal, meaning "I like you because you're so strong and good to me." When accompanied by the rear of the body held low and broad tail swings, this is a very submissive gesture, which contains the element of "I'm no threat and I don't want you to hurt me."

This same flattened ear gesture, accompanied by a relaxed open mouth, blinking eyes, and a reasonably high tail, is a sign of friendliness. It means, "Hi there. We can have fun together." This set of gestures is usually followed by a clear invitation to play, such as a play-bow or a stutter-bark.

Ears pulled slightly back to give the impression of a slightly splayed or sideward spread of the ears: To me, the heads of prick-eared dogs have always seemed to be describable as a sort of a V-shape in which the top of the V is the ears and the bottom point of the V is the muzzle. The ear signal we are considering now has the effect of widening or opening out the V. In some animals, the ears may appear to flatten out slightly to

RELAXED / ATTENTIVE

DOMINANT / AGGRESSIVE

FEARFUL / AGGRESSIVE

FEARFUL / SUBMISSIVE

Figure 9-1   Basic ear positions of a prick-eared dog.

the side, taking on a shape something like airplane wings, but not quite that flat. This is a very ambivalent sign. Basically it means, "I don't like this," and, "I'm ready to fight or to run." This position of the ears indicates that the animal may quickly turn from uneasy suspicion to aggression, or to fear and escape behaviors.

Ears flickering, usually slightly forward, and then a moment or so later slightly back or downward: Another sign of indecision, but with a much more submissive and fearful component. It could be read as saying, "I'm just looking this situation over, so please don't take offense." In this sense, it has a much greater pacifying content. Once, in an obedience class, a dog trainer and I were watching this changing pattern of ear signals in a Siberian husky named Eddie. As his ears flicked forward, back, and occasionally sideways, the trainer laughed and she said, "When Eddie starts doing that with his ears, I always feel he's just trying on different emotions to see which one fits the current situation."

LOP EARS AND LOPPED EARS

Adults of the wild canine species, whether wolf, jackal, coyote, dingo, fox, or wild dog, all have pricked ears; but their puppies all have floppy ears that hang down like flaps beside their heads. Only in some domestic breeds do the lop ears continue on in adults. This leads us to ask how this "juvenilization" of the ears came about in these dog breeds. Since all of the ear signals we've mentioned seem to have been designed for pricked ears, we must also ask whether ear communication is somehow affected in adult dogs with lop ears.

It is important to remember that the dog breeds we know today were created by a primitive sort of applied behavior genetics. There are certain characteristics which people seem to have wanted from all of their dog breeds, and these seem to have been consciously selected for during the process of domestication of dogs. Specifically, people wanted dogs who were relatively tame and willing to accept direction and control by those they see as leaders. Genetic manipulations are never simple. When you select for one characteristic you want, you often find it genetically linked to other characteristics that may be desirable or undesirable. For instance, the same set of genes that determines the white coat color in dogs can also carry a predisposition toward deafness. Breeding for tameness (which is behavior characteristic of wild puppies) also produces dogs that are physically more like puppies, with shorter muzzles, less developed teeth, bigger eyes, smaller and rounder heads, and, most importantly for us here, the lop ears found in some breeds.

Originally, ear shape was not an important characteristic for most dog breeds. It did not affect a dog's ability to hunt, track, retrieve, or herd. For this reason, people breeding for function and purpose paid little attention to the occurrence of lop ears. Ear shape does affect the animal's look, however. To many people, the look of long floppy ears is quite appealing, perhaps because it is reminiscent of the effects of long hair framing a human face, or perhaps simply because it retains something of the puppylike look in the adult.

With the rise of "dog fancy" as a sport, where dogs are judged solely upon their looks, ear shape became an important issue. The Norwich terrier was created in the 1880s by Frank ("Roughrider") Jones. He used working terriers from several English kennels, probably leaning heavily on a mix of border terriers, cairn terriers, and Irish terriers. Eventually, he created a fine small terrier that could work alone or in packs to dispatch foxes and rodents. However, Norwich terriers came with two separate ear types: a lop-ear and a pricked-ear type. Though this was of no consequence to the early owners of the breed, it later became important to the dog fancy. It was eventually understood that ear type did breed true, and matings of pricked-eared dogs produced pricked-eared dogs, while matings of lop-eared dogs produced lop-eared dogs. So, in 1979, the American Kennel Club separated the breeds, allowing the prick-eared variety to continue with the label of Norwich terrier while renaming the lop-eared variety a Norfolk terrier. I don't know what rationale was given for each name, but it was fortunate for me with my often imperfect memory. I can now use a simple memory device to distinguish the two: "Norwich" is similar to "Nor-witch," and traditional witches apparel includes pointy hats reminiscent of the pointy ears on the Norwich terrier.

When it comes to dog communication, lop-ears are not a trivial matter. Pricked ears provide more visible signals than lop-ears do. Changes in ear shape are more easily seen and at longer distance, making communication more positive and less ambiguous in prick-eared dogs. This is not to say that one can't read ear position in dogs with floppy ears, but rather that the signals of lop-eared dogs are more subtle, both for human and canine observers.

Take a look at Figure 9-2, which shows examples of the various ear positions in a lop-eared dog. The top left example shows the ear position of a relaxed and attentive dog. To the right is the lop-eared dog's equivalent to ears up and forward in a prick-eared dog. This particular look can mean increasing dominance and possibly aggression when accompanied by appropriate facial and body language. This look always re-

RELAXED   DOMINANT / MILD AGGRESSION

SUBMISSIVE

Figure 9-2   Some basic ear position signals in a lop-eared dog.

minds me of a full frontal view of an elephant, with its ears held out to the sides. The bottom figure shows a more submissive ear position, and is the lop-eared equivalent of pointed ears pulled back along the head. To me, this look appears as if the dog's ears have been pulled downward and the ear flaps pasted against the side of the head. Notice, then, that although the signals are less pronounced, there are still variations in position which allow lop-eared dogs to communicate their feelings and intentions.

There are some odd and sad aspects to this story about dog's ears. As we have seen, breeders genetically created the races of lop-eared dogs and thus decreased the visibility of important communication signals. But, never content with what they have, people now step in again. After producing lop-ears, the breeders have decided to surgically alter many of these dogs by lopping off these very same long ears. This practice of ear cropping involves cutting off much of the external ear. It is usually only done to dogs that have floppy ears in the first place, and the effect of this is to even further impede the communications ability of the these dogs.

Several reasons are given as to why certain breeds of dogs, such as boxers, Doberman Pinchers, Rottweilers, or Great Danes, have much of their external ear flap cut off. The breeds that have their ears cropped were originally designed to be guard dogs. In all dogs, their ears and attachment areas of the ears are extremely sensitive, and when damaged, cause extreme pain. Leaving the lopped ears on a guard dog places him at risk. They provide intruders with a pair of handles they can use to grab the dog by, while still managing to stay clear of his teeth. Not only does the culprit control the dog's head by grasping the dog's ears, he also causes the dog incredible pain. Both of these problems can be eliminated in one fell swoop by simply cropping the ears back, so that only short stubs remain, which are too small to effectively grab and hold.

Recently, ear cropping has become the focus of an international controversy. Some countries have banned the practice and others are considering such a ban. Proponents of cropping have come up with some new rationalizations to support the practice. One is that certain breeds are "hearing breeds" (meaning that hearing is important for them to perform their normal functions well) and require maximum sensitivity for the use of their ears. The large, pendulous ear flap covers the auditory canal, and thus obviously reduces the amount of sound reaching the dog's inner ear. Supposedly, removal of the external ear allows direct access of the sound to the auditory canal, hence making the dog more sensitive to sounds.

Another argument is that ear cropping promotes hygiene, since dogs with long floppy ears are prone to ear infections. Ear flaps can keep moisture in the ears from drying out and this can promote ear infections and other problems.

These arguments do not stand up well to close scrutiny. Many dogs, such as hounds, spaniels, and retrievers, have much longer and thicker ears than the working breeds that usually have their ears cropped. Nobody seems to argue that these breeds should have their ears cut off, even though many of the hunting and retrieving breeds are trained to work to whistle-signal commands and certainly this should classify them as "hearing breeds." As to the ear hygiene argument, it is much more likely that a spaniel or a retriever, rather than a boxer or a Rottweiler, will be forced to work in water, and thus be exposed to potential ear infections. Yet breeders of these sporting dogs don't seem to have a problem with lop-ears, and none has suggested that such ears should be cropped to improve hearing or for hygienic reasons.

Aside from any pain the surgical procedure might inflict upon dogs, cropping also affects the dog's communication ability. In theory, it should be possible to find a method of cropping that replaces the lopped ear with something approximating a pricked ear, and if the musculature is appropriate, this could actually improve communication through ear talk. Something like the very long fancy "show crop" that is sometimes used for Doberman pinschers comes closest to this. However, this show crop is difficult and expensive, and defeats the purpose of cropping in a guard dog by still leaving too much ear to grab. The standard, utility, and guard-style ear cropping that you see on most dogs of this breed leave too little of the ear to be visible, and not enough of the rear section to create visible changes of ear position in response to muscle movements. This suggests that breeds of dogs whose communication abilities have already been reduced because of the genetic changes which gave them lop ears, now undergo an even more drastic reduction of signaling ability due to surgical alteration. Except for working guard dogs, it seems to me that ear cropping today is mostly a matter of cosmetics—searching for a particular style or "look."

Unfortunately, I have to offer my opinions on this matter as merely opinions. I know of no scientific research on the communication ability of dogs with cropped versus uncropped ears. But I can offer one set of actual observations which may bear on the subject. One acquaintance of mine has two handsome, neutered male boxers, Zero and Naught. At the time, both were around three years of age, and both were quite socialized and friendly, as boxers often can be. The major visible difference between them was that one of them had cropped ears and the other did not. On an outing, it was interesting to watch the two of them as they wandered around interacting with other canine guests. Zero, the one with cropped ears, was much more likely to be approached with suspicion by other dogs. Unfamiliar dogs seemed to treat the upright stubs of his ears as if they were the challenge signal of a dominant animal, and would often stiffen as they approached and act cautiously in their greeting behaviors. Naught, with his natural lop ears, was more likely to be approached without hesitation by the other dogs.

There are many possible explanations for these different responses, which may have nothing to do with cropped ears. However, the most obvious ones did not seem to apply here. Generally speaking, larger dogs are perceived as more dominant and potentially threatening than smaller dogs. But for a male boxer, Zero was on the smallish side, standing around 23 inches. Naught was at least 10 pounds heavier, and stood almost 2 inches taller at the shoulder. On the basis of size and appearance, then, Naught, should have appeared to be more threatening, and I would have expected more caution and hesitation when other dogs approached him. In addition, Zero was actually the more passive and submissive of the two, making the actions of the other dogs even more unusual. Although I have no scientific evidence to support the conclusion, I have the feeling that the somewhat cool response given to Zero was because his ear signals were so difficult to read and the shape of his cropped ears was easily confused with the upright, forward ear signal of a dominance threat. Also, the fact that there was so little readable change in Zero's cropped ear positions could have been read as a failure to acknowledge the friendly gestures from approaching dogs, and this could be misinterpreted as a warning of possible aggression.

I suppose I'm left with a general feeling that we should leave the lop-eared dogs alone. Their signaling ability is already moderately hampered by their lop ears, so let them use whatever abilities they have left to communicate as best they can. There is really no sound reason for us to intervene surgically.

Actually, thinking about dog's use of their ears has caused me to reconsider my earlier statement that human beings never use their ears for communication purposes. People certainly pay attention to each other's ears, which is why we ornament them with rings, studs, and pendants. It is in this aspect of human behavior that communication is possible, but in a more deliberate and relatively static way. I once had a girlfriend who had two sets of gold earrings, each with a word that hung on a tiny chain from the stud. One set said, "Yes"; the other said, "No." She used to tell me: "If you want to know where you stand with me, just read my ears."

One evening I came to pick her up to take her out to dinner. My spirits brightened for the evening when I noticed a "Yes" hanging from her ear. I was chattering away in my most cheerful manner as I helped her to slip into her coat. It was at that moment that I noticed that from her other ear hung a golden "No." The messages sent by most dogs' ears, even those of cropped and lop-eared dogs, are more dynamic and certainly less ambiguous than the one she was sending.

10

Eye Talk

For animals that live on land, all faces follow the same master plan, whether you are a snake, toad, lion, human, or dog. Faces were originally designed to control the quest for food, and since everything that an animal can put into its mouth is not necessarily edible or safe, the three major screening senses—taste, smell, and sight—are clustered near the mouth. The arrangement is always the same: taste buds in the mouth, the nostrils of the nose just above, and the eyes perched just a bit higher. This arrangement allows land animals to eat bits of food on the ground, while sniffing smells and viewing what it is eating all at the same time.

The general pattern of the face is the same for most species, but there are some differences in the eyes. Animals that are preyed upon, and whose only defense is to run, need an early warning system. For this reason, rabbits and antelope have their eyes on the sides of their heads, to give them a full panoramic view of the world, sometimes covering the full 360 degrees. This makes it very difficult for anything to sneak up on them. In predators, such as tigers or wolves, the eyes are placed facing front, like headlights. This gives the opportunity for binocular vision, which sharpens the ability to read the distance of things. Obviously, knowing how far away something is in order to pounce upon it accurately would make an animal a better hunter. Dogs are predators, so their eyes face forward as well.

The eyes, however, have more than a visual function. As human beings, we feel that the most expressive part of the body is the face, and one of the most expressive parts of the face is the eyes. Actors and film directors know the importance of eyes as a means of communication and often take advantage of it. The classic thriller director Alfred Hitchcock said, "Dialogue should simply be a sound among other sounds, just something that comes out of the mouths of people whose eyes tell the story in visual terms." He often used close-up shots in which virtually only the eyes were visible, to give a sense of menace or fear. Actor Henry

Fonda felt that important messages were sent by the eyes and always insisted on having a "catchlight" for his close-ups. In the film trade this is called an "inky-dink," and it is a tiny light which is placed near the face. When an actor looks directly into it, his or her eyes glow and appear to have a deep emotional intensity.

In dogs, there are several important structural aspects of the eyes which can provide interpretable communication. The colored part of the eye is the *iris;* the hole or dark spot in its center the *pupil.* The white portion of the eye is called the *sclera;* it is the outer covering of the eye. Finally, the visible shape of the eye is determined by the way in which the eyelids open or close.

PUPIL TALK

When it comes to actual vision, the sole purpose of the iris, or colored portion of the eye, is to contract or expand to vary the size of the pupil and thus control the amount of light actually getting into the eye. In dim light, the pupil expands to gather in whatever light energy might be around, while in bright light, it contracts to prevent too much glare from washing out details of the visual scene. Pupils, however, communicate. The size of the pupil, and its dynamic changes in shape, also can come about because of emotional states.

Generally speaking, excitement, interest, or any intense emotion will expand the pupils. There are many studies which have looked at what causes pupils to expand in humans. Interest in another person is certainly one such factor. Though people are not often consciously aware of the size of another person's pupils, we do pick up that there is something about their eyes that seems to suggest pleasure or interest. Since we tend to view people who are interested in us more positively, someone who looks at us with large, dilated pupils is apt to appear more attractive to us. Beginning in the Renaissance era and up through the nineteenth century, women used poisonous extracts from belladonna (also known as the deadly nightshade plant) to dilate the pupils of their eyes to make them appear more attractive to men. The very name of the plant, *Belladonna,* is Italian for "beautiful lady." In today's world, we accomplish much the same thing when we have a candlelit dinner. The dim light makes the pupils expand (to let in more light); thus we look more attractive and interested without the administrations of a toxic substance. One particularly interesting aspect of the attraction of large pupils is that we have specifically bred some of the toy dogs for large eyes and large pupils. These are mostly companion dogs. The large pupils of dogs like the Cavalier King Charles spaniel or the Pekingese appear to radiate affection to the human observer.

Just as in people, the size of a dog's pupils also reflect its emotional state. The difficulty is that for dogs, pupil size is sometimes harder to read, since some breeds of dogs have very dark irises and the pupil may seem to blend into the dark surround. The lighter the irises, the easier it is to see the changes in pupil size. Even in dark eyes it is worth looking carefully, since the eyes speak quite loudly about the dog's feelings.

If large pupils indicate an intense emotion, then smaller pupils will often indicate boredom, drowsiness, and relaxation. It is important to remember that changes in the size of a dog's pupils only reflect changes in the *intensity* of an emotion, not necessarily whether that emotion is positive or negative. Great joy and excitement can result in wide pupils, but so can great fear or anger. However, if you happen to look into a dog's eyes at the critical moment when you can actually see the pupils in the process of expanding and contracting (rather than just the final position of a large or small opening), you can get additional information. A joyful or a pleasant arousing situation results in a simple widening of the pupil. When the animal is becoming angry or aggressive, the pupils start to change size by first contracting and then expanding to their larger aroused size.

DIRECTION OF GAZE

Let's turn next to the white region, or sclera, of the eye. Surprisingly, this also serves a communication function. Why should evolution have made part of the eye white? Why not simply extend the color of the iris to make the whole eye brown or blue? The reason is that the white of the eye contrasts with the color of the iris, and this makes it easier to see which direction the eyes are gazing. In humans, this is an extremely important method of communication, and for this reason we have a very large expanse of white compared to other animals. Dogs also have a white sclera, but sometimes you have to look carefully to pick up cues from it. Part of the reason for this is that dogs often turn their head in the same direction that they turn their eyes.

In people, the ability to detect the direction that others are looking is socially important. It tells us who others are listening to in a conversation. It signals their intentions, since individuals tend to look in the direction they are about to move. Master salesmen say they can tell which items a person is most interested in by following the line of their gaze. When the potential customer's eyes start to flick in the direction of the exit door, they know the customer has lost interest or is bored or uncomfortable and thinking about leaving.

Looking at an individual can communicate a great deal and sometimes can trigger behaviors in other people. I remember reading an article about a recent violent confrontation between two British gangs following a very close soccer game. People were badly hurt and parts of the stadium were damaged before police finally came and stopped the battle. The strange thing about this clash is that it all started with when one gang member pointed his finger at an opposing gang member and started to yell: "He looked at me! Did you see that? He looked at me!" The columnist who was describing this incident went on to make a large point about how deranged the people who belong to such gangs must be to begin a battle simply because someone was looking at them.

The truth of the matter is that looking at someone is not an innocuous event. Staring is definitely viewed as a threat. Psychologists have conducted some interesting real-world experiments to prove this.[1] In one study, researchers stood on a street corner and gazed steadily at drivers who had stopped at a red light. They found that the vast majority of drivers noticed the stare within seconds, and when the light finally turned green, they raced across the intersection much faster than drivers who had not been stared at. In other studies, the researchers stared at pedestrians. In response, they walked faster to try to move away from the staring individual. In yet another study, the researchers had their associates stare at students in a university library, and found that the students who were stared at were more likely to pack up their things and leave the library earlier.

Virtually all animals view staring as a threat. This goes all the way down the evolutionary scale to reptiles. The hog-nosed snake will fake death if a potential predator comes near. If the predator is staring at it, it will fake death for a longer time. You can freeze lizards in place by staring at them, or get a defensive reaction out of many birds by staring. Monkeys seldom ignore being stared at, and will respond either submissively or aggressively to such behavior. Dogs also use staring as a controlling gesture. Let's look at this and some other eye signals.

A direct eye-to-eye stare: A wide-eyed direct stare is often a threat, an expression of dominance, or even an announcement that an attack is about to begin. Frequently, a dominant dog or wolf will approach a less dominant animal and directly stare at them. The animal that is lower in dominance will usually break off eye contact, turn away, or even lie down in a submissive gesture. If the direct stare is not responded to, an escalation of the level of confrontation may follow. Thus, this expression is best translated as something like "I'm boss around here, so back off" or "You're annoying me. Stop that or I will make you regret your behavior."

There is an interesting way in which dogs use the direct stare to control human behavior. It is most frequently seen at the dinner table, or when people are sitting around eating something. The dog comes over and sits staring at a person and then at the food they're eating. This is an obvious attempt to get some food, and it is most likely to work when the dog is small. In this case, even the most direct of stares is less apt to be read as a threat by the human who is on the other end of the gaze. The person might interpret this look as being "doleful," "hopeful," or "pleading," and respond by giving the dog a morsel of food. From the dog's viewpoint, however, they are directly asserting dominance with that stare. When you respond by giving him what he wants, the dog interprets this as a submissive gesture on your part, and also reads this as your acceptance that the dog has a higher status in the pack than you do. This is a dangerous precedent to set with a large dog and a potentially troublesome one even with a small dog. It will ultimately make the dog more difficult to control, since you must be a leader, or at least higher up on the dominance ladder, if you are to be obeyed. This is a clear case where you should think about what the dog is saying before you give any response.

On the other hand, you should be careful about staring directly at an unfamiliar dog. Staring at a dominant dog may be viewed as aggression, while staring at a fearful dog confirms its worst fears and may result in panic on its part. But staring can be used quite successfully to control your own dog. Staring at the dog directly can often make him stop a particular behavior, and many dogs will respond with submissive and pacifying gestures to try to win back your favor.

Eyes turned away to avoid direct eye contact: If a direct stare is a threat, then it seems logical that breaking off eye contact would be a sign of submission, and perhaps of fear. This is certainly true for dogs. A dog will avert it eyes when confronted by a dominant dog. It will typically look down and away in a movement that is best interpreted as reading, "I accept the fact that you're the boss," and, "I don't want any trouble."

The same holds true for humans. I'm sure you've heard mothers instructing their children, "It's rude to stare at someone." In casual conversation, we usually avoid directly staring into a person's eyes. Instead, we let our eyes dart around their faces. We glance to the side, look into our coffee cup, or only look directly at them when they are looking away.

Avoiding direct eye contact with someone of authority is almost a ritual. An average person would never make direct eye contact with the great priest or the emperor. Remember the nursery rhyme line, "Only a cat can look at the king"? That's the way it is with canines. When the pack leader returns, the others all gather around him, glancing at his face, but never staring directly into his eyes.

The eyes turned away can have other meanings in dogs as well. In certain situations, this gesture can mean boredom. One sometimes sees this in dog obedience classes when there are long pauses between exercises. The dogs who had been watching their masters begin to break off any eye contact and look around in a rather bland way. When attention wavers, the gaze begins to drift aimlessly.

Blinking: Most animals blink. Even the most attentive human being loses about twenty-three minutes of visual information from his or her waking day because of blinking. These minutes disappear in 14,000 tiny gaps in the flow of vision. Blinking is a necessary activity, however. Eyes must be kept moist and clean, and the cells of our cornea (the clear bulge in the front of our eyes) must be kept alive. Each blink sweeps the fluid from our tear glands over the eyeball. Tears are not just water but part of the circulatory system. Because the cornea must be transparent, there can't be any blood vessels running through it, so one of the jobs that tears have is to carry oxygen and nutrients to keep the corneal cells alive. Tears also contain a variety of chemicals that kill bacteria and trap dust and debris. Blinks move the tear fluid over the eye. About three-quarters of the tears eventually leave the eye and seep down into the nasal passages. This helps to keep the nose moist and bacteria-free. It also explains why crying makes your nose runny.

The rate at which we blink, and the situations in which we blink (or not), can be sources of information about emotional states. High rates of blinking can show boredom; we blink less when we are paying attention. When drivers have been at the wheel for a long time, their blink rate increases. If something interesting appears near the road, they blink considerably less. Of more importance is the fact that blinking can also serve as a sign of submission. When describing a conflict, it is common to refer to the person who backed away from the confrontation as the one who "blinked first." We refer to the tough and confident individual as the one who can make difficult decisions and carry them through "without batting an eye."

In the language of dogs, blinking breaks the dominance stare and shows submission. While it does represent a giving up of dominance, it is not as submissive a gesture as full-scale aversion of the eyes. The eye blink is more apt to mean, "We are almost equals, but I will accept your leadership," rather than, "Please don't hurt me and I will follow your direction."

Blinking also can be a sign of friendliness or even attraction. We all know the stereotype of the coy maiden who bats her eyes when a man looks at her in a gentle invitation. For dogs and wolves, blinking can actually be part of a greeting ritual. When a submissive dog approaches the pack leader or another dominant dog, it lowers its body slightly and may lick at the air or the face of the dominant animal. If this approach is accepted, the dominant animal will often blink two or three times in rapid succession. The submissive dog then blinks in return, and may continue to lick the air or make swallowing or chewing movements with its mouth. In effect, the pair has now agreed to accept each other on friendly terms.

EYE SHAPE

The shape of a dog's eyes is easy to make out because most dog breeds have contrasting colors at the rim of the eyes. In lighter dogs, this tends to look almost as if a dark outline of the eye has been penciled in. For darker dogs, there is often a slight lightening on the face in the eye regions or on the fur or membranes surrounding the eyes. The purpose of these marks is simply to allow the shape of the eye to be seen at a longer distance.

The language of eye shape is quite simple: the larger and rounder the visible eye, the more angry and threatening the dog. Wide eyes are part of the pattern of dominance staring. In fact, the muscles beneath the eyes pull in (as part of the pattern of movements that cause the angry wrinkling of the dog's nose and forehead), and this puts pressure on the eye to force it slightly forward in its socket. As a result, first, more of the eye is exposed, making it appear to be even larger. Second, the increased visible surface area of the eye will now more easily catch any available light, making it more prominent and easily seen.

The reverse muscle action makes the eye become smaller, less visible, and narrower, and these changes are associated with fear, submission, and pacification. A dog who is really trying to put off a threat and to show an extreme level of submission may actually even close its eyes.

There is one situation where this eye-shape language may break down. This is the case where fear and aggression are combined, as when a lower-ranking dog has been forced into a situation from which it cannot escape and feels that it must fight. Under these circumstances, the eye may take on something like a teardrop or triangular shape, slightly wider near the nose and noticeably narrower toward the temples, as if the eye were trying to express two emotions. It begins with a wide stare near the nose and gradually tapers so that the eyelid hoods the eye and makes it look small. The conflicted state of the dog's mind clearly shows in the conflicting message in its eyes.

For humans, much of the message conveyed by the eyes comes from the changing pattern of the eyebrows. Since they contrast in color from our skin, they are readily visible from a distance. Small muscle movements around the eyes and the forehead, which might otherwise be difficult to see, are accentuated by the presence of visible eyebrows.

Eyebrows send clear messages, and we read them accurately and automatically. Scientific studies have shown that if you screen off all of the lower face, leaving only the top portion containing the eyebrows and the forehead, people can still reliably identify most of the basic human emotions. There are even certain expressions which almost solely involve the eyebrows, such as the so-called brow-flash response. This response occurs when we greet a friend from a distance away. It involves a quick raising and lowering of the eyebrow, which lasts about a sixth of a second. The response indicates friendliness and pleasure, and it does not happen when we greet people that we don't like. Although people seem not to be consciously aware of this signal, it appears to be universal and is found in Europeans, Americans, Samoans, the Bushmen of Africa, and even isolated tribes in Peru.

Dogs don't have eyebrows. Humans need eyebrows, since their function is to keep perspiration from dripping down our foreheads and into our eyes. Dog don't need eyebrows because they don't sweat in the same way that we do. The only place where dogs sweat is from the pads of their feet. However, dogs do have evolutionary precursors to our eyebrows in the form of markings that accentuate the movements of the muscles around the eyes. These can assist in communication. In many dogs, different fur colors show up as spots near the eyes. According to folk tradition, the dogs with the highest degree of psychic ability are the "four-eyed dogs," which are light-colored dogs with a dark spot over each eye, or black dogs with light spots over each eye. These dogs are supposed to have special psychic abilities, such as the ability to see demons, devils, or ghosts. While I can't attest to their mystical talents, it is likely that they obtained this reputation because their expressions were easier to read than those of other dogs. The contrasting-colored spots make the movements of the muscles over the eye much more visible.

For some dog breeds which lack such spots, the pigmentation that frames the eye will extend beyond the eye's border to mark a region on the brow. For some other breeds of dogs, especially uniformly colored and dark dogs, the pattern of fur growing around the upper portion of the eyes changes to produce a characteristic shadow pattern that serves the same function as the spots on the four-eyed dogs. All of these marking differences simply serve to allow us to read the dog's expressions "as if" it had eyebrows.

The eyebrow (or eye-marking) language of the dog is similar to the eyebrow language of people. When the dog becomes angry, the space between the brows (spots) is contracted and the brows angle downward. The hair will also rise up around the eyes to give a stronger impression of a human-style eyebrow. When the dog is afraid, or being submissive, the brows are pulled upward in the center, and the outside ends are pulled down and out toward the temples. This movement is less dramatic because the hair around the eyes remains quite flat, and there is no accentuation of the eyebrow region.

Dogs also use their eyebrows to show perplexity and concentration, usually when they are working on a problem or trying to interpret something. This shows up as a pull downward and together, without the angled slant found in anger. It is the same movement that humans make when thinking hard about something. For this reason, Charles Darwin called the *corrugator* (the muscle which controls this movement) "the muscle of difficulty."

Dogs can show fairly subtle emotions with their eyebrows. Surprise and even mild wonderment are signaled by an upward and outward movements of the brows. I even knew one Airedale, named Brandon, who had a mischievous eyebrow expression. He would look at you and then raise one eyebrow. This gesture was always followed by his grabbing something and then making a dancing dash away with it, in an attempt to elicit a "catch me if you can" game.

WEEPING

All mammals have the apparatus to make tears, but they use it solely to keep their eyes moist and clean. It is often claimed that only human beings actually weep as an emotional expression. For humans, the emotions that produce tears are usually negative, such as grief or pain, but if a positive emotion is strong enough, we may "weep for joy." Recently, several researchers have suggested that many mammals weep, but only under conditions of extreme emotion.

While attending a meeting of the Canadian Veterinary Medicine Association a few years ago, I started quite an argument over the issue of whether dogs weep. There were about a dozen of us sitting at a dinner table, eight of whom were veterinarians. When I asked whether dogs weep, these vets split down the middle, with four of them arguing that dogs do cry as we do, and four claiming that such tears were only reflex in nature, due to muscle tension in the face and around the eyes caused by pain and fear. Both sides became quite excited (and a bit loud) as they tried to make their points. Obviously, this shows that the issue is controversial. In my own experience, I have seen a dog weep once, and have indirect evidence that another dog wept.

One late afternoon, I was walking across my university campus near a construction site. Suddenly, I heard the most frantic and piteous screaming, which sounded like a human baby in dire agony. I raced over, to find a young female boxer (I would later learn she was called Evita) struggling with some barbed wire. She must have been caught on the strand of wire and probably had spun around a number of times trying to get free. The effect of that spinning was to wrap several loops of the wire around her, and her struggles were causing the barbs to cut deeply into her sides, back, and belly. I flung off my coat and threw it over her so that I could control her struggling without cutting myself. At that same moment, a workman who had stayed late on the site dashed over. He saw what was happening and grabbed some wire cutters. While I held Evita, he cut the wire strands surrounding her. All the while I was making soft sounds and watching her face, which was dominated by two great dark eyes from which trickles of tears wended their way down her cheeks. To me, it made sense. She was in great pain, and she was crying as a hurting and frightened child might cry.

The second time I saw evidence of weeping in a dog I was too late to see the actual crying or to help. My old cairn terrier, Flint, had been in pain during the night, and although he seemed better in the morning, I decided to take him to the vet to be checked out. When I arrived at his office, I opened the back door of the van, only to find that Flint had quietly died sometime during the twenty-minute ride. I had heard him whimper plaintively during the ride and had tried to reassure him that we would soon have help for him. I looked at that old grizzled face and noticed tear tracks running down his muzzle. There was no doubt in my mind at that moment that he had been weeping. I also knew that there were matching tear tracks running down my face.

11

Tail Talk

People who don't understand the language that dogs speak can often get themselves and others into trouble. One day, I got a phone call from Steve, a Professor of Education I had met at several university functions. He sounded very upset.

"I need your help," he said. "My dog has suddenly turned into some kind of a sociopath. He attacks without warning, and last night he bit my grandson. Now my daughter says she won't bring the child back to our house unless we do something about the dog. She thinks he should be destroyed. She says that any dog that bites a child should be destroyed. He really isn't a bad dog. Could you do something for him?"

Early that evening, I went to Steve's home, and he met me at the door. Standing behind him was a beagle, who sized up the situation, gave a greeting bark, and wiggled his way past his master in a friendly manner. As I bent to rub his chest and ears, Steve informed me, "That's Bagel. Normally he's just like this—as friendly as can be. Recently, though, he's become unpredictable. He's bitten me, my wife, and now Denny, my grandson. I just don't know what to do. We love him, but we won't live with a dangerous or psychotic dog. If we have to we'll . . ." Steve's voice trailed off and he looked sadly at the dog.

I was quite surprised by all of this. Beagles have a few problems, such as a strong tendency to wander off with their nose glued to the ground, completely ignoring their owner's frantic shouts. They can be a bit hardheaded and independent, at least when it comes to learning and obeying obedience commands such as "Sit," "Come," or "Down." They are also easily distracted by things going on around them. Some people also complain about their tendency to "sing" or bay in a high tenor tone when they are excited. However, beagles have always maintained their popularity as pets because they are so affectionate and have a low level of aggression. In a house full of active kids they will play quite happily, while in a home with elderly people they are just as happy adopting the role of couch ornament. In addition, they are generally known to be tranquil and tolerant. From my first encounter with Bagel, he seemed no different than any others of his breed, so I was quite puzzled.

Steve went on giving me details as I knelt and fondled Bagel. "We've provided him with lots of toys. You know, rubber things that bounce and can be chewed, fuzzy toys with squeakers inside, and dozens of those rawhide chew things. He usually picks one of them and then jumps up on the couch to chew on it. That's what he was doing when I brought Denny over to pat him. You know, it was really strange. He stood up and looked so happy to meet my grandson, but when Denny stuck his hand out, he gave a growl and bit him without any warning at all!"

I was now a bit puzzled. "Steve, when he bit your wife, what were the circumstances?"

"Nothing special. Actually, it was sort of similar. He was on the couch with another toy, and when my wife went over, he stood up in a happy way to say hello, and then he growled and bit her."

I thought I knew what was happening. "Steve, I'm sure you've heard that some dogs can feel very threatened when they're chewing on something or eating something and then a person comes close. If that person reaches their hand out to pat them, it's possible the dog might misinterpret this as an attempt to steal the chewy object. Stealing food or a cherished object from a dog can often cause the dog to respond aggressively."

Steve looked at me, and in a tone that professors of education use when they're trying to explain a difficult concept to a rather dumb student, he said, "Of course I know that. As a start, you should know that Bagel is not that kind of dog. That's why I asked you to come look at him to find out what is wrong. Second, I think I can read a dog as well as the next person. If he looked threatened, if he snarled or whimpered, I wouldn't try to touch him, and I certainly wouldn't send my four-year-old grandson to him. But that isn't what happened. He stood up. He looked right at him—right into his eyes! All the while he's wagging his tail. Then, when Denny went to pet him, he growled and bit him!"

"Okay. Just *how* did he wag his tail?"

Steve's voice now took on the tone of an annoyed professor faced with a student who is disturbing the rest of the class with his stupid and unproductive questions. "I think that's obvious. It's the way that a dog moves his tail from side to side to show you he's happy." Steve was moving his arm back and forth in front of my face to give this stupid student a visual demonstration of the concept.

"Steve, just a minute more. Was his tail sort of low or horizontal, making big swings back and forth that shook his hips as well?"

Now Steve was looking puzzled. He squinted his eyes a little as though he were trying to look at a brightly projected memory of the incident. "No, it wasn't like that."

"Did he hold his tail high?" Steve nodded. "Was it almost vertical and wagging in short movements, more like vibrating rather than swishing or waving?" I demonstrated with my hand and Steve again nodded.

The rest of my task was really quite easy. It was a matter of explaining to Steve that all tail wags do not signify the same thing. Some tail wags are indeed associated with happiness. However, others can mean a variety of different things, ranging from fear and insecurity, through dominance challenging, all the way up to a clear warning that if you continue your approach, you are apt to be bitten.

In Bagel's case, he was indeed protecting his chew toy. He was not wagging his tail in happiness to greet the approaching person. Rather, when he was approached, he stood up, made a direct wide-eyed threat stare, and held his tail in a high, dominant, threatening manner. That tail was saying, "Back away! I will protect what is mine!" When Steve's grandson apparently ignored that warning, Bagel felt justified in carrying through his threat in the only way that he knew—by biting. Understanding Bagel's tail talk was the first step toward solving the problem.

In some ways, tail-wagging serves the same functions as our human smile, polite greeting, or nod of recognition. Smiles are social signals, and human beings seem to reserve most of their smiles for social situations, where somebody is around to see them. Sometimes, vicarious social situations, as when watching television or occasionally when thinking about somebody special, can trigger a smile. For dogs, the tail wag seems to have the same properties. A dog will wag its tail for a person or another dog. It may wag its tail for a cat, horse, mouse, or perhaps even a butterfly. But when the dog is by itself, it will not wag its tail to any lifeless thing. If you put a bowl of food down, the dog will wag its tail to express its gratitude to you. In contrast, when the dog walks into a room and finds its bowl full, it will approach and eat the food just as happily, but with no tail-wagging other than perhaps a slight excitement tremor. This is one indication that tail-wagging is meant as communication or language. In the same way that we don't talk to walls, dogs don't wag their tails to things that are not apparently alive and socially responsive.

A dog's tail speaks volumes about his mental state, his social position, and his intentions. How the tail came to be a communication device is an interesting story.

The dog's tail was originally designed to assist the dog in its balance. When a dog is running and has to turn quickly, it throws the front part of its body in the direction it wants to go. Its back then bends, but its forward velocity is such that the hindquarters will tend to continue in the original direction. Left unchecked, this movement might result in the dog's rear swinging widely, which could greatly slow its rate of movement or even cause the dog to topple over as it tries to make a high-speed turn. The dog's tail helps to prevent this. Throwing the tail in the same direction that the body is turning serves as a sort of counterweight, which reduces the tendency to spin off course. Dogs will also use their tails when walking along narrow surfaces. By deliberately swinging the tail to one side or the other in the direction opposite to any tilt in the body, the dog helps maintain its balance, much the same way a circus tightrope walker uses a balance bar. Quite obviously, then, the tail has important uses associated with specific movements. However, the tail is not particularly important on flat surfaces, when a dog is simply standing around or walking at normal speeds. At these times, it becomes available for other uses. Evolution again seized an opportunity and now adapted the tail for communication purposes.

It is something of a surprise to many people to learn that puppies don't wag their tails when they are very young. The youngest puppy I ever saw systematically wagging its tail was eighteen days old, and both the breeder and I agreed that this was quite unusual. Although there are some differences among the various breeds, the scientific data suggests that, on average, by thirty days of age, about half of all puppies are tail-wagging, and the behavior is usually fully established by around forty-nine days of age.

Why does it take so long for the puppy to start wagging its tail? The answer comes from the fact that puppies begin wagging their tails when it is necessary for purposes of social communication. Until they are about three weeks of age, puppies mostly eat and sleep. They are not interacting significantly with their littermates other than curling up together to keep warm as they sleep or crowding together to nurse. They are physically capable of wagging their tails at this time, but they don't.

By the age of six or seven weeks (when we start to see tail-wagging behaviors on a regular basis), the puppies are socially interacting with one another. Most of the social interactions in puppies consist of what psychologists call "play behaviors." It is through playing that puppies learn about their own abilities, how they can interact with their environment, and most important, how to get along with other individuals. A puppy learns that if it bites a littermate, it is apt to be bitten back, and perhaps the game it was playing might be terminated by its now angry playmate. It is at this point that the puppy also starts to learn dog language. It is not clear to what degree these emerging social communications are prewired, but learning is clearly needed to refine the use and interpretation of these signals. The pups learn to connect their own signals and the signals provided by their mother and their siblings with the behaviors that come next. They also begin to learn that they can use signals to indicate their intentions and to circumvent any conflicts. This is where and when the tail-wagging behavior begins.

One place where conflicts are likely to occur is during feeding. When a puppy wants to suckle its mother, it must come very close to its littermates as it crowds in to find her teats. Remember that this puppy is now coming close to the very same individuals that might have been nipping, jostling, or chasing him a few minutes earlier. To indicate that this is a peaceful situation, and to calm any fearful or aggressive response by the other puppies when they too are pushing toward the mother's teat, the puppy begins to wag its tail. Tail-wagging in the puppy then serves as a truce flag to its littermates. Later on, puppies will begin to wag their tails when they are begging food from the adult animals in their pack or family. The puppies come close, to lick the face of the adult, and they signal their peaceful intentions by tail-wagging. It thus becomes clear that the reason that very young puppies don't wag their tails is that they don't yet need to send appeasement signals to other dogs. When communication between dogs *is* needed, they rapidly learn the appropriate tail signals.

Tail language actually has three different channels of information: position, shape, and movement. Movement is a very important aspect of the signal, since dog's eyes are much more sensitive to movement than they are to details or colors. This makes a waving or wagging tail very visible to other dogs.

Evolution has used a few additional tricks to make the tails even more visible. Wild canines, like wolves, often have great bushy tails, which are easily seen at a distance. In addition, many tails are specially colored to facilitate recognition of tail signals. Often, the underside of the tail is lighter, to make the high-tailed signals quite visibly different from signals involving the tucking of tails into a lower position. Many canines will also have distinctive markings to make the tail tip more visible. Usually, there is a lightening toward the tail tip, or perhaps simply a white mark which defines the tip of the tail. In other canines, the tail tip is noticeably darker. Either of these two color contrasts helps to make the end of the tail more visible, and this make movement and position cues easier to recognize.

TAIL POSITION

Although I will begin here by describing tail position, remember that each signal sent by the tail can have up to three different components. As with other signals, it is the combination of the various components which produces the richest and easiest to read meanings. There is also another important factor to be considered. Different dog breeds carry their tails at different heights. All tail positions should be read relative to the average position that a dog normally holds its tail. We'll return to this problem later.

Tail horizontal, pointing away from the dog but not stiff: This is a sign of attention. It roughly translates to "Something interesting may be happening here." It is usually triggered by some event occurring nearby or the approach of someone at a distance. It is sometimes associated with an interesting smell carried in the wind. There is no sign of threat in this gesture, but if the tail begins to stiffen, it indicates that the dog feels that the situation is beginning to change to something else.

Tail horizontally straight out, stiff, and pointing away from the dog: Stiff tails usually contain an element of aggression; thus, this is part of an initial challenge when meeting a stranger or intruder. It roughly translates to "Let's establish who's boss here," and is the beginning of a rather cautious greeting ritual between dogs that don't know each other well. It will also sometimes be seen if an incident has caused some competition, such as two dogs finding a bit of food or a prize toy at the same time. Since it is the leader of the pack, or the most dominant dog, who gets first choice of any food or treasures, the outcome of the challenge is important. Such an exchange seldom leads to any physical aggression, since normally one of the two dogs will size things up and (perhaps as a result of a previous history in confrontations) may simply back away, thus resolving the conflict.

The position of the tail can have a major effect on the meaning of the gesture. Let's modify the height of the stiff tail by raising it a bit.

Tail up, between the horizontal and vertical position: This is the sign of a dominant dog. The stiffness of the tail indicates the dog's intention is to actively assert his dominance over anybody near. The dog does not feel challenged at the moment, but anticipates the possibility that he might be challenged. This tail signal translates to "I'm boss around here, and I'm willing to prove it to anyone who doubts my word."

Removing the stiffness from the tail, but still holding that high position and moving the tip of the tail forward a bit, shows a definitely confident attitude.

Tail up and slightly curved over the back: This signal says, "I'm top dog, and everybody knows it." This is the expression of a confident, dominant dog, who feels in control and has no doubts about it. This dog expects no challenges; everything it wants will happen according to its plans and desires.

I often wondered how these high tail positions indicating the dominance of a particular dog evolved. Strangely enough, I was given a clue that started me thinking along the correct path by a folk tale. It happened when the Dalai Lama was visiting Vancouver to give a series of speeches and to make some political contacts. The Dalai Lama is the spiritual head of the largest order of Tibetan Buddhists, and until 1959, he was also the political ruler of Tibet. The university held a special reception following a public address, and I had been invited to attend. There was a lot of security around the Lama, and the room was so crowded with dignitaries that I never did actually get to meet him. However, a number of Buddhist monks from his entourage were also present, and I did get the opportunity to chat for a while with one of them.

My interests had nothing to do with politics or religion, but, as always, with dogs. In this case, it had to do with the Lhasa Apso, which is the oldest and the most popular of the four breeds of dogs considered native to Tibet. This is a toy breed, which stands about 10 inches at the shoulder and weighs around 14 pounds. It has a long, silky coat, drooping ears, a small tail held high, and a flattened face. In theory, it was supposed to look something like the Celestial Lion. Its history goes back at least 1,300 years. It has a long association with the Buddhist monasteries, where it was kept as a companion dog and also as a watchdog to sound the alarm. There was also a tradition of bringing this little dog into the room of a dying priest. The monks believed that the dog would serve as a temporary home for the soul of the holy man until he could be reincarnated in a new human body. Because of this association with the souls of sanctified men, they were prized pets. Over the years, successive Dalai Lamas gave Lhasa Apsos as gifts to several Chinese emperors. I hoped I might learn something new about these little dogs by speaking to someone from their homeland.

I found one of the Lama's attendant priests who spoke very fluent English, and seemed friendly and willing to talk. He laughed when I asked about the dog, probably because everyone else had politics and other weighty matters on their minds. Nonetheless, he did tell me about them.

"These dogs became very popular in the seventeenth century. It was during the time of the fifth Dalai Lama, whom we call the Great Fifth [Ngag-dbang rgya-mtsho]. He was a military and political leader, and he formed an alliance with the Mongols. [He actually is responsible for the political dominance of this religious order in Tibet.] The Great Fifth was also very aware of popular traditions and stories, and he would often tell stories for amusement. One story that he told had to do with the tail on the Lhasa Apso. The Mongols and the Chinese often used war dogs in their campaigns. According to the Lama, the gods gave their dogs their tails to serve the same function as the flags carried by military leaders. The upright flag told the fighting troops where their commander was located, so they could look there for instructions or rally around that place for defense. The Great Fifth said that this is the reason the dog who is the leader carries his tail erect. It is his flag. The Lama would usually laugh when he said that he always felt that only fighting dogs had really earned their tails. For this reason, he would sometimes suggest that Lhasas should probably have been born tailless. In later years, his mind changed. This happened sometime after he had built the Potala, in Lhasa. [This is the place that would later give the dog its name. The Potala is the beautiful winter palace used by Dalai Lamas.]

"The Great Fifth had many political enemies who wished him dead. It is said that one night, while the Dalai Lama was sleeping, assassins stole into his quarters in the palace. They silently killed the group of guards on the outer perimeter, then stealthily approached the set of interior guards who were actually guarding the Lama's own bedroom. Suddenly, a great loud barking broke out from the little Lhasa Apso who slept in the Lama's chambers. This alerted the Lama's personal guards and caused others to come from nearby. Once having been detected, there was no chance that the assassins could succeed. It was in this way—as a warrior guard—that the Lhasa Apso saved the life of the Dalai Lama. Afterwards, the Great Fifth was overheard telling his dog, 'Now I understand that you have truly earned your tail, little dog. Carry your battle flag high, and with great honor.'"

I've always thought this a charming story, and sometimes recall it when I see a dog with its tail erect. Much later, I was viewing a series of wolf films, some of which involved wolves greeting each other after a hunt, or preparing for a hunt, when I realized that there was a kernel of truth in the idea that the upright tail was a dog's battle flag. In these episodes, the group of wolves would often crowd around the pack leader. In that mass of milling canines it was often difficult to locate any one individual, except for the pack leader. His tail was indeed quite high, like a flag, so that his location was always unambiguous.

The idea of a flag to rally the pack becomes even clearer when the effect of a high tail is considered in different situations. I noticed, for example, that when the pack leader wandered around the resting area with his tail relaxed, the individual pack members paid only casual attention to his movements and would continue with their activities. If, however, the pack leader moved across the clearing with his tail upright, pack members were much more likely to take notice, and more than that, to actually move to the leader's side. The Alpha wolf seemed to use this tail signal selectively. He raised his tail when preparing to gather his comrades to begin a hunt. He also raised his tail when approaching an unfamiliar animal, or when taking notice of an ambiguous or possibly threatening situation. The raising of the tail gathered his troops around him, much as the raised flag of the Mongol leader served as a marker for his forces to gather around him.

As the tail lowers, the content of the message changes.

Tail held lower than the horizontal but still some distance off from the legs, perhaps with an occasional relaxed swishing back and forth: This is a signal from an unconcerned dog with no particular worries at the moment. It translates to "I'm relaxed" or "All is well."

Tail down, near hind legs: This tail position modifies its meaning depending upon the body language of the dog. If the legs are still straight, and the tail swings back and forth slowly, over a fairly short distance, we can interpret this message as meaning, "I'm not feeling well." This is a common signal for a dog who is ill or in a moderate degree of pain. The signal can also refer to mental rather than to physical distress, and in that case it could be translated as "I'm a bit depressed."

With a change in the body position there is a change in the meaning of this signal. The most common modification involves a lowering of the body, by bending the hind legs slightly inward. This gives a slight downward slope to the dog's back, and it changes the meaning to one containing some elements of apprehension or timidity. Basically, this tail sign now means, "I'm feeling a bit insecure." It is most frequently seen when the dog is introduced to an unfamiliar environment, but also sometimes occurs when the dog sees a member of its family leaving the house and anticipates a period of separation from its usual companions.

Tail tucked between the legs: Lowering the tail fully moves the signal from apprehension or mental discomfort to fear. This position translates as "I'm frightened!" or "Don't hurt me!"

While fear is the major element in this tail position, it also has become ritualized to a pacifying signal to help fend off aggression from another dog. The most common circumstance in which this tail position occurs is in the presence of a dominant dog or a person who is perceived as being dominant and in control. In these instances, this tail signal can also mean, "I accept my lowly role in the pack, and I'm not trying to challenge you," or, "I'm obviously too overwhelmed by you ever to doubt your position of authority."

There is an interesting additional reason why high tail positions may have evolved as dominance signals while low tail positions may have become signals of submission or insecurity. This reason has nothing to do with the visual signal that the tail provides. The argument is that the important thing is not necessarily the tail itself, but rather, what is beneath it. The anal glands of a dog carry a lot of scent information which serves to identify the animal, and also gives some indication about its emotional state and sexual receptivity. The anal glands are virtually a written report describing personal information about a particular dog. A dog who raises its tail thus exposes this information to the world. It advertises the dog's identity by making its scents available to everybody nearby. This could be seen as the equivalent of putting one's name up in lights or having one's biography published for all to read. In people, such public exposure of personal information is associated with individuals who are famous, rich, powerful, or important for other reasons, and hence truly confident. The same holds true with dogs. The dominant dogs are proud of their identity and are pleased to broadcast to everyone, "Your leader has arrived. Take a sniff so you know who I am."

If the high tail position is designed to expose the anal glands and release scent, then obviously a lower tail position will tend to reduce the amount of scent. A tail firmly tucked between the dog's legs will physically cover the anal region, effectively serving the same function as the stopper or cap on a bottle of perfume, physically preventing any odors from escaping. In essence, the dog is making its presence less obvious by preventing the release of the odors that identify it as an individual. Some scientists have suggested that the tail-between-the-legs action is the canine equivalent of an action seen in insecure human beings—especially children—who hide their faces when brought into the presence of a dominant or potentially threatening person. Thus scent signals may also be an important component of tail talk.

Figure 11-1 gives a visual image of the relative tail positions. Notice that with increasing dominance or aggression, the tail position rises, while with increasing fear or submissiveness, the tail position lowers.

TAIL SHAPE

As I noted earlier, the information from the position of the tail is moderated by several factors. One of these is the tail's shape.

Bristling hair down the tail: The easiest way for a dog to change the

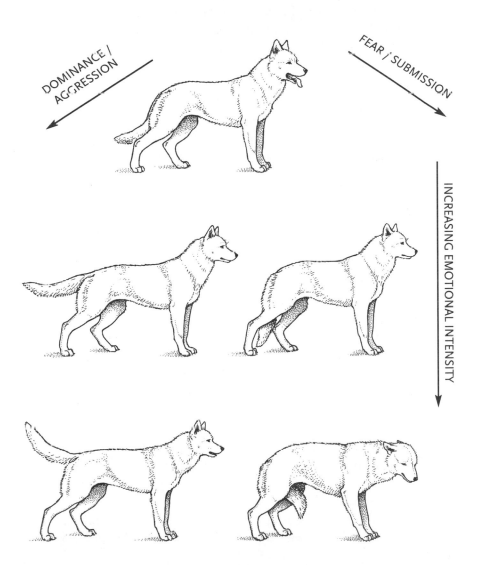

Figure 11-1 Basic tail positions: The top figure represents a relaxed, alert dog. Moving down the left column shows that the tail raises with increasing dominance or aggressive threat, while moving down the right column shows that with increasing fear or submission, the tail position lowers.

shape of its tail is by causing the fur to rise up, instead of lying flat as it usually does. Generally speaking, the brain center which causes the hair on the dog's hackles to rise also causes the hair on the tail to rise. Like the hair standing up on the dog's shoulders, the bristle in the dog's tail is a sign of aggression. It is, however, an independent signal which can be used to modify any tail position by adding a bit of threat to the signal. Thus, with the tail straight out, it modifies the signal from "Let's establish who is boss around here" to become "We're going to determine who is boss around here and I'm ready to fight if you think that's needed to settle the issue!" Adding the bristle to a tail that is held up or over the back, it means, "I'm the leader here. I'm not afraid of you, and any challenge from you will result in a fight." The tail held at a lower level, while bristling, means, "You're making me nervous and anxious; if you push me, I may be forced to fight."

Tail bristling only at the tip: While the full tail bristling always carries an aggressive message, a tail that bristles only at the tip, especially if that tip is raised, adds a component of either fear, anxiety, or despondency, rather than aggression, to the message. Thus, a tail held down (but not tucked between the legs) with a bristle at the end and the tip bent a bit up, is apt to be saying, "I've got a case of the blues today." In my dogs, this is usually cured by giving them a bit more personal attention. If that fails, I start looking for a physical problem that may be making the dog unhappy.

A crick or sharp bend in the tail when the tail is held high: There is an interesting change in tail shape that occurs most visibly in the dogs that are the most wolflike in appearance. German shepherd dogs, Belgian sheepdogs, and some of the northern breeds show this most prominently. This signal looks like a crick or a sharp bend in the tail. It will sometimes give the tail a bent or broken look, or make it appear to have a snakelike, rippling look, sort of like a letter S lying on its side. It is a definite sign that immediate aggression is being contemplated by the dog. If you are faced with this signal, especially if there are other signs of dominance or aggression, it is time to remove yourself and your dog from the situation. This sign translates: "Back off! If you don't get out of here, I'm going to attack—now!"

A crick near the tip of the tail: This adds a moderately aggressive threat to any other signal. It says, "Stand back. You don't want to push me or I may attack."

Tail cricks have to be looked for. They are often a bit subtle but should be responded to, since they do often suggest that the dog is feeling aggressive enough to actually bite.

TAIL MOVEMENTS

The various movements of the tail can add additional shadings and meanings to many other messages, whether they are conveyed by sound, body language, or other aspects of tail talk.

Fast tail-wagging: This tail-wagging can come about as a sign of excitement or tension. Generally speaking, the vigor or speed of the wag indicates the degree of excitement. It is important in judging excitement from tail movements to attend to the speed of wagging, waving, or vibration, independent of the size of sweep of the movement. The size of the tail wags will differ depending upon breed characteristics, so you should look carefully. For example, a sporting dog with a full-flowing tail might seem to move its tail much more than a terrier moves its carrot-shaped tail (which may actually only seem to tremor). Yet, in both cases, high-speed movements simply mean the dog is excited. The size of each tail sweep tells us whether the dog's emotional state is positive or negative rather than about the dog's level of excitement.

Slight tail wag, each swing of only small size: This tail wag is usually seen during greetings. It can directed toward strangers, or to the human master or another significant family member when they enter the house. It often occurs before the dog's presence has actually been apparently noticed by the person. It is best interpreted as a tentative "Hello there," or a hopeful "I'm here."

This small tail wag can also be seen when the dog's owner shifts his or her gaze toward the dog, either when entering a room or as an interruption in some ongoing activity. It is often given with the interpretation of "I see you looking at me. You like me, don't you?" This is a response to social attention, and is designed as reassurance that the dog is not challenging the person's position of dominance, but rather is seeking social comfort and friendly support.

Broad tail wag: This is a friendly, "I'm not challenging or threatening you," signal. It can also be used to say, "I like you." You will often see this tail gesture during play, when one dog seems to be attacking the other, pouncing, growling, and barking. At the same time that all of these apparently aggressive noises and behaviors are being produced, the tail is wagging broadly. It is this wag that is designed to be a reassurance to the other dog (or person) that all of those other signals are in fun—like kids playing cops and robbers where one pretends to shoot the other and does so with a great angry roar, "I got you this time! Now I'm gonna kill you!" accompanied by a huge smile.

This broad tail wag can also mean, "I'm pleased," in many contexts, and this is the closest to the popular conception of the "happiness" tail wag.

Broad tail wag, with wide swings that actually pull the hips from side to side: This is sometimes seen in greeting, especially after a long absence. It may appear when the particular dog's "special person"—the one whom the dog seems to listen to and obey the best—appears in the room. It may also occur if the dog is learning a new command, such as the recall, where the dog learns to respond to the word "Come." Although this appears to be the happiest of all tail wags, it is actually a complex signal, indicating the social rank of the person relative to the dog. You can interpret this as meaning something really quite effusive and humble, sort of like "Oh, Great Leader, I am here for you. I will do what you want me to do and you, in turn, will care for me and not harm me."

When the dog is really trying to lay on the "Great Leader" compliments and the "I will follow you anywhere" message, it will lower the rear part of its body so that the tail swings actually seem to be almost sweeping the floor. At the same time, the front end of the body rises ever so slightly, the dog looks up imploringly, and it licks at us or simply licks the air in front of us. At the sight of such love and respect, we read the message as true and accept the dog, pet it, greet it, and feel protective toward it. This is exactly the behavior that lower-ranking dogs show to the leader of the pack when it returns or approaches. The message is respect, non-threatening compliance, and pacification of any potential aggression.

Slow tail wag, with tail at "half-mast": This is less social than most of the other tail signals. During dog training, I interpret it as "I'm trying to understand you. I want to know what you mean, but I just can't quite figure it out." When the dog finally solves the problem, the speed and size of the tail wags will usually markedly increase, perhaps to the point of becoming the large "Great Leader, I hear and obey" wagging.

Generally speaking, slow tail wags, with the tail neither in a particularly dominant (high) nor subservient (low) position, are signs of insecurity or of being unsure as to what to do next. You will sometimes see this kind of wag when a dog notices someone approaching its home or territory. He may take a step or two toward the stranger, give a slow tail wag, then look back toward his family or pack, followed by another slow tail wag, another look at the stranger, and so forth. The indecision is reflected in his tail. The moment the dog decides whether this is a danger, threat, or positive event, or the moment he decides on a course of action, his tail will rise or fall, and this "indecision wag" will be replaced by another, more obvious signal.

HUMAN MEDDLING AND TAIL DIALECTS

The actual form that tail signals can take in any given dog varies because of the tail shape and normal carrying position of any particular breed. Human intervention and establishment of breed standards by the dog fancy have dictated the "standard" for tails. Certain breeds are required to carry their tails low, while other breeds must carry theirs high, and still others must carry their tails at midlevel. Dogs with too high or too low a natural position for their breed are usually penalized in competition. Some breeds must have a tail that is naturally straight; others must have a tail that naturally curls over the body; others again must have tails that are usually tucked between the legs. Some breeds are required to have a plumed tail with many "feathers," while others are forbidden plumes. Some breeds must have tails of specific lengths, while others must be tailless.

Many but not all of these requirements are just for show or in order to obtain a particular "look." However, each breed of dog was bred to perform a specific function, and in some cases the tail is an essential part of that performance. This is particularly true in hunting dogs. Setters were designed to move over the terrain much more quickly than their predecessors, the pointers. They were also bred so that the hunter could tell how near his dogs felt that they were to the game by the beat of their tail. Specifically, their tails beat faster and faster as they get closer to the game. These tails are supposed to be well feathered to make them visible. Once the setters actually locate the game, all tail movements are supposed to stop as the dog "points" to its quarry. The cessation of tail movements tells the hunter that the birds are very close to the dog. Thus, the hunter knows he must approach carefully lest he spook them from their hiding place before he is near enough to take an effective shot at them.

The northern sled dogs are supposed to hold their tails high and, once again, there are functional reasons for this tail position. The high carriage of the tails on sled dogs makes any signals visible to the driver, even if the dogs are harnessed to the sled. Deviations from the normal high position can be seen even when the sled is moving. When all tails are up, the team is attentive and ready to go. A relative droop in a dog's tail will be easily visible, and the driver can determine whether that particular animal has a problem. The straightening of a tail, or the development of a visible crick, tells the driver there may be conflict among the various team members. If the tails were held at an intermediate level it would be much more difficult to see these signals, since the tails of the leading dogs would be blocked by the bodies of the dogs hitched behind them. Thus, without the high tail positioning, the driver would be deprived of vital information about his team.

Herding dogs are supposed to carry their tails low. Usually, their tails are relatively still, and point backwards, following the same direction as the dog's body. The reason for this is that herding dogs use nips, stares, and charges in particular directions to move the herd. The direction taken by the herd is usually determined by the dog's posture. Sheep, for instance, will move directly away from a staring dog, or run quickly in a straight line away from a charging dog. The animals in the herd seem to use the alignment of the dog's head and body as though it were an arrow, pointing in the direction they are supposed to move. Imagine if the herding dog had an upright, waving tail, like the sled dog. This kind of signal could easily distract the herd animals, pulling their attention away from the dog's direction of gaze and alignment, and obscuring the message the herding dog is trying to send.

The problem is that the "breed-required" tail shape and positions can sometimes blur messages sent to humans or other dogs. People looking at an Irish setter, for example, are apt to read the rapid beat of its tail as an indication that the dog is enthusiastic, or in some cases even overexuberant. One seldom hears of an Irish setter being referred to as withdrawn or too reserved. On the other hand, you often do hear such comments made about dogs like border collies, whose tails lie low and don't move with much enthusiasm. Yet I have found border collies to be as sociable as most setters. The difference is that, because of the breeding for specific tail attributes, one must learn to read the signals in the context of the specific breed. Of course, if you live with an individual dog for a while, it should become easier to read its relative changes in tail position, and its tail talk becomes less ambiguous.

Perhaps the most significant way in which human beings have interfered with the normal tail signals in dogs is through the practice of tail docking, in which part or all of the tail is amputated at birth. Obviously, a dog without a tail cannot give tail signals.

There is a lot of heated controversy about tail docking, and I have mixed feelings about the issue. The arguments against it (based upon suggestions of cruelty, pain and mutilation) have caused several countries to ban this practice. Nonetheless, it is important to understand the original purposes that caused dog breeders to adopt it. Docking did not begin simply as a matter of fashion, where breeders strive for a particular look in the show ring. Many breeds of spaniels who have their tails routinely docked have quite elegant and well-plumed tails, which actually make the dog appear more handsome, at least to my eye. As in many cases of human manipulation of dog form, the practice of tail docking began for very practical reasons.

One reason is the same as the reason for originally cropping the ears of guard dogs. The rationale is that the tail provides a way whereby a criminal or other undesirable person could grab hold of the dog, controlling its actions and thus avoiding its teeth. The tail is docked fairly closely to prevent the use of such strategies against guard dogs.

However, most of the dogs who have their tails docked are not guard dogs. Partial or full tail docking actually occurs in over fifty breeds of dogs. In many of the sporting breeds, it was originally meant as a preventative procedure to avoid tail damage. Such damage is particularly common in those hunting breeds which have to pursue game through heavy vegetation, brambles, or over rocky terrain. The fast action of their tails, as they whip back and forth, can easily lead to a torn, broken, and bleeding tail, which is painful, often difficult to treat, and may require the riskier amputation of the tail in the adult dog. Obviously, cutting off the tail eliminates the risk of injury.

The rationale for this procedure has recently been confirmed in a study conducted by the Swedish German Shorthaired Pointer Breed Council. After a ban on tail docking was introduced in Sweden in 1989, there was a noticeable increase in the number of tail injuries reported by the breed. In 1991, this organization looked at 191 of these pointers who had undocked tails and at the time of the study were between twenty-four and thirty months of age. An amazing 51 percent of these dogs had sustained tail injuries requiring some form of medical treatment. The likelihood and severity of these injuries appears to be linked to some fairly obvious factors. The researchers mentioned the liveliness of the dog and its tail motions as one such factor. As expected, how often the dog was used to hunt, and the type of terrain, were also important factors. Dogs used in bushy, woody, or rocky terrain were much more likely to suffer tail injuries than those used in marshy areas or on level grassy regions.

Dogs with thicker, well-muscled tails, like Labrador retrievers, do not have their tails docked and seem less likely to suffer injuries. In some breeds, such as Vizslas, the lower part of the tail is quite strong, while the section of the tail nearer the tip is often turned upward (making it more likely to snag on obstacles) and carries little fat and muscle as protection from brush and rocks. For this reason, only the upper third of the tail is usually docked in this breed.

I can understand the practicality of these reasons for tail docking, but

I still worry that docking significantly limits the use of tail signals and hence reduces the effectiveness of a major channel of communication in dogs. Let me offer a bit of data and an anecdote to support this. In one study, we observed dogs interacting in a confined city park area where they were allowed to be off leash. We tallied 431 encounters between dogs. Most of these (382, or 88 percent) were typical canine greeting behaviors, often followed by play behaviors, including the usual chase games. The remaining forty-nine encounters contained an aggressive element on the part of one or more of the dogs involved. These could be as mild as a snarl and a snap with no physical contact or as severe as an actual physical assault drawing blood.

The dogs involved were coded simply on the basis of whether they were tailless (most likely docked) or with tail (undocked or partially docked). To be classified as tailless, the dog had to have a tail approximately 6 inches or less in length (we eliminated small toy dogs from the sample). The proportion of dogs with tails was considerably higher in this population, and amounted to 76 percent, as opposed to 24 percent without tails. However, when we looked at the dogs involved in aggressive incidents, twenty-six (or 53 percent) of these confrontations included dogs without tails. On the basis of the number of dogs with and without tails, we would have expected only twelve aggressive incidents (24 percent) to involve tailless dogs. Thus, our results show that dogs with short or absent tails are twice as likely to have aggressive encounters than dogs with longer visible tails. One cannot help but wonder if the increase in aggressive encounters might not be related to the ambiguity or absence of appropriate tail signals which could have indicated pacification and thus avoided a confrontation.

The anecdote about tail docking has to do with a Labrador retriever named Transit. He was a typical Lab, with a friendly attitude toward life. He was kissy-faced around people, and interacted well with dogs. His owner, Mark, often took him to a nearby park that permits off-leash dogs in a fenced area. Mark reports that Transit never had any negative or aggressive interactions with any of the local dogs there. Then, one day, this happy black dog was involved in a freak accident involving a garage door with an automatic closing device. He suffered a severely mangled tail. When Transit was rushed off to the veterinarian, the only course of action was to amputate the tail almost completely, leaving a stub of about an inch or so. Transit recovered completely, and, with the resilience that continues to keep Labradors in the top ten breeds in terms of popularity, his personality showed no change at all—at least as far as humans are able to read. Mark, however, reports that other dogs now respond to Transit in a more ambiguous manner. When he meets new dogs, it seems to take longer for them to complete the greeting ritual, and three times in the three months since Transit had healed well enough to return to the park, there have been snapping or biting incidents. In every case, the attack was initiated by the other dog. Could it be that Transit has lost some of the precision in his communication? Could it be that when he lost his ability to tail-talk, he could no longer send the clear pacifying and friendly signals he used to?

While these data and this anecdote could have other explanations, and these may involve factors other than tail docking, they have caused me to wonder. I accept the "preventive safety" reasons for tail docking, but worry that the communication deficit, which seems to be a consequence, is significant. Perhaps it is time to seek a compromise in this controversy. Perhaps it is time to consider partial docking—removal of that portion of the tail (usually toward the tip) which is most likely to be injured, while leaving enough of the tail to allow the dog to communicate with its canine associates. Unfortunately, I doubt that this idea will be accepted. The opponents of docking can still maintain that removal of any portion of the tail is mutilation of the dog, while the advocates can maintain that we are not removing enough of the tail to make these working dogs fully safe.

Another solution to the problem would involve a lot more time and creative work. Why not breed sporting dogs with stronger tails and guard dogs with shorter tails? It could be done. We've bred dogs for all sorts of similar attributes. It might require some relaxation of some breed standards to allow a certain degree of outcrossing to effect the changes, but could well be worth the effort. However, I fear that the dog fancy will not permit such "adulterations" to occur, which means that the likelihood of this solution being accepted is minimal. So, if I had a tail which had not been docked by evolutionary pressures, you would find it now carried quite low, near my legs, with little movement: an obvious indication of my emotional state over the tail-docking issue.

12

Body Talk

I was attending a Municipal Park Board meeting where there was a debate going on about a regulation that made it illegal to have a dog off leash in any park. There was some public controversy over the recommendation to designate some park areas where dogs could run free (under appropriate owner supervision). Several people presented their views for and against this change in regulations and sometimes the discussions became pretty heated. At one point a woman was speaking, giving an impassioned argument against the off-leash areas, claiming that dogs were "messy and dangerous." The man sitting next to me was a university professor who taught negotiating skills in the Faculty of Commerce. He glanced over at me and remarked, "The new regulation will win by three, maybe four votes. We'll get our off-leash parks."

"How do you know that?" I asked.

"They told me," was his reply, as he gestured toward the board members sitting at a long table in the front of the room.

"You mean you talked to them before this meeting?"

"No, they're telling everybody. Look at the guy on the right. See the way he's leaning forward to listen? He agrees with this speaker, and so does that woman two seats down stroking her chin. Now look at the others. None of them are in favor of her argument. Two are leaning back as if trying to increase their distance from her, and one is looking at the ceiling. The man next to him has his arms folded over his chest, the woman beside him has her hands locked in front of her and her teeth clenched. That woman with her finger against her face is also rejecting this argument. The only vote I can't be sure of is the guy with the beard. The way he leans his head against his hand indicates boredom, and he may go either way, although from the angle of his body I'll guess he thinks what she's saying is pretty dumb."

He had read the body language of the people quite accurately. The idea of a trial period with a limited set of parks designated for off-leash recreation with dogs was passed by four votes. Even though these public officials were not deliberately trying to give their opinions in advance, their bodies gave accurate signals about their attitudes and intentions.

Professional negotiators, clinical psychologists, some law enforcement agents, and many business people all are trained to read the nonverbal signals that body language provides. Most of us have become pretty good at doing the same thing even though we haven't been specifically schooled for this skill. Take a look at Figure 12-1, which shows a few stick figures. Now look at this list of statements and try to determine which figure is most likely to be saying each of the following phrases. Write the letter corresponding to that figure next to the phrase. When you have finished, go on to the next paragraph for the correct answers.

Figure 12-1 These stick figures are a test of your ability to read body language. Write the letter corresponding to each figure next to the appropriate statement in the text, then check the following paragraph to see how well you did.

_____1. "We won!"

_____2. "I really don't want any of that."

_____3. "Welcome to my home."

_____4. "Don't you forget I'm still boss around here!"

_____5. "This is all very embarrassing to me."

_____6. "Let me think about this."

_____7. "I didn't know anything about what was going on here."

_____8. "I'm very upset and disappointed in you."

The correct matches are: 1-E; 2-A; 3-C; 4-F; 5-B; 6-H; 7-D; 8-G. I'm sure most of you interpreted most of these instances of body language with very little difficulty even though you may have never had any formal training to do so. Notice that the information conveyed in these brief "snapshots" of people actually deals with a number of complex issues, such as greeting, social dominance, anger or annoyance, excitement, innocence, and so forth. All of this information is communicated by a person's posture, hand positions, the way they carry their heads, and the way they move.

The same thing is true for dogs. They use body position, paw placements, and the way they move as a vital part of their language. Also, like people, their body language sends messages about their emotional states and social matters.

BASIC BODY LANGUAGE

There is a general rule which goes with the expression of social dominance, aggression, fear, and submission in body language. It is that the more aggressive and dominant the dog is, the larger and taller it makes itself appear to be. Submissive, frightened animals try to make themselves small. This is not a new observation; Charles Darwin noticed and described it back in 1872 in his book *The Expression of the Emotions in Animals and Man.* Let's see how this general principle combines with the dog's other body movements to send some very specific messages.

Stiff-legged, upright posture or slow, stiff-legged movement forward: This is the body language of a dominant dog who is trying to say, "I'm in charge around here." It also contains a definite suggestion that, if necessary, physical aggression will be used to assert authority. Thus, another meaning of this posture is "I challenge you." This posture is shown in the original Darwin drawing, which I have included as Figure 12-2.

For a long time, people thought that this posture indicated the dog was getting ready to fight and that aggression was unavoidable. This is far from true. Dominant dogs seldom actually fight because they don't need to do so. The rank ordering in a pack of wild canines, such as wolves, is normally established without any bloodshed. The threat of aggression is really an example of ritualized behavior, where the signs and signals are important, not the actual actions they seem to prepare for. The word "ritual" comes from the Latin word *ritualis*, which means "habit" or "ceremony." These are behavior patterns which have lost their original function as a preparation for an action and have instead gained meaning for communication purposes. Even though the threat is unlikely to be followed by a real attack, the effect that the signaled threat produces in other animals in the group is usually sufficient to establish the dog's rank order in the pack.

Why has this body gesture become a signal and part of the dog's com-

Figure 12-2   A dog asserting dominance, as shown by Charles Darwin, *The Expression of the Emotions in Animals and Man,* 1872.

munication pattern, rather than simply being the first stage of a true attack? The answer has to do with evolution and survival. Think of it this way. Many minor encounters occur every day which might result in some form of conflict. They include disputes over who gets to sleep where, who moves out of whose way, who gets to eat first, who gets to initiate play or sex, and so forth. If each of these daily situations led to a physical fight, the canines would all eventually become drained of energy, and would spend much time nursing their wounds and healing. This would certainly reduce the chances of survival for both the individual and the whole pack, since a tired, injured animal can neither hunt efficiently nor defend itself effectively.

Evolution plays its role here. The dogs that have learned to accept the dominance signal and submit to it are better off in terms of health and energy. The dogs that have learned to give a dominance display and wait for compliance, thus not needing to actually follow up with a fight, are similarly better off. Because of this, all individuals will be more likely to live on and to reproduce. This means that evolution actually favors less physically aggressive individuals in groups of social animals, while at the same time allowing some real fear-provoking displays to remain as part of the canine communication system.

If two dogs both assume this upright posture, and if they recognize that they are approximately equal in dominance but not a threat to each other, a little greeting dance begins. The dogs blink or break eye contact momentarily, then slowly move toward each other's flanks, avoiding any further direct staring. When the two are standing next to each other, both still with their tails in the air, they sniff each other's anal region. This serves two purposes. It helps each one to recognize the other's sex and identity, and it also signals that each one is confident enough to expose itself without fear of attack. After this, they may circle around each other a few times, then both may dash off together to play or go their separate ways.

This does not mean that this full dominance display is *never* followed by a real attack. Since we are looking at communication, once the signal is sent, the sender's next action depends upon what the other dog does in response.

Body slightly sloped forward, feet braced: This is the signal most likely to provoke a follow-through attack. The dog in this body position has seen a dominant dog's statement that he is boss, but is not accepting this statement. In effect, he is saying, "I'm challenging your dominance and am ready to fight!" At this point, virtually anything can happen. The encounter might end peacefully if the first dog backs off now, or at least stops trying to assert its dominance. This suggests that he accepts the second dog as at least his equal. If not, both dogs may continue to move toward each other and may then actually fight it out.

Small changes in the signals at this point can also tell you what is about to happen next. In this case, you should watch for the hair on the dogs' backs.

Hair bristles on back and shoulders: This is a sign of possible aggression even when it occurs without the stiff upright body posture. A ridge of hair bristling down the back says, "Don't push me!" or, "I'm getting angry!" Under some other circumstances, it can also indicate fear and uncertainty.

It is important to look carefully at the pattern of bristling. In many breeds, the individual hairs darken toward the tips. Thus, when the dog raises its hackles (extending the hairs down shoulders and spine), the dark tips render them more visible. This makes the dog look even larger and taller, reinforcing an expression of dominance. In some wolves, and some dog breeds, there is a noticeable line or patch of dark hair down the back, and occasionally a darkening at the shoulders. Presumably these color changes are designed to attract the eye to these signals.

There are two ways in which the hackles can be raised. One of them involves hair bristling only in the region of the neck and shoulders. A dominant dog, who is still confident and only moderately troubled by the current situation, is likely only to raise the hairs on these upper-back regions. The second way involves extending the hairs all the way along the back (which may be accompanied by the tail bristling as well). This "full display" of bristling means, "I've had it with you," and is a sign of imminent attack. In other circumstances, it may also mean that the dog is a bit uncertain and worried about the situation and is preparing to use its teeth in a desperate effort to defend itself or its position in the pack. In any event, whether rising anger or rising fear, this pattern of hair bristling often indicates that the dog sees no alternative other than actually fighting, unless the other dog backs off or gives way.

Lowering the body or cringing, while looking up: This is a clearly submissive gesture, since it involves lowering the body to make the animal look small—the opposite of expressing dominance by making itself look larger. The dog is basically saying, "Let's not argue," or, "I accept your leadership and higher social status." Darwin caught the classic form of this posture in Figure 12-3.

Some people have suggested that this is the emotional expression of a dog who is simply afraid of the animal or person it is lowering itself in front of. Fear, however, comes in several flavors. The most obvious form might be called "existential fear," where the dog's very life and safety are threatened. In this situation, there are really only two courses of action open to the dog. It can escape the fearful situation, or it can fight the individual who is frightening it. Most sensible dogs will run if they are frightened enough. Escape is the best course of action, since it minimizes the chance that the dog will be hurt.

The narrow hips on most dogs (best seen in the sight hounds like the greyhound or whippet) give them good running speed, and this seems like a good option when threatened. If, however, escape is impossible, as when a dog is cornered by a larger predator such as a bear or a mountain lion, then the only option left to preserve life is to fight—at least long enough to find an escape route. Let us imagine that a dog is confronted by a 400-pound grizzly bear and there is no way to run. Do you really think the dog would assume a posture such as we see in Figure 12-3? Of course not. This would do no good, and would only make it easier for the bear to hurt him.

There is a second form of fear, however, which we might call "social fear." This results when social animals, such as dogs, come into conflict with members of their own species. Obviously, escape and fighting are

Figure 12-3   A dog showing submission, as shown by Charles Darwin, *The Expression of the Emotions in Animals and Man,* 1872.

still options open to the dog. Of the two, fighting is a less likely outcome. Remember, evolution seems to have taken a dislike to aggression among members of the same group, unless there is no other recourse. Escape is always a possibility. Running away from an animal of much higher status will reduce anxiety, but at the same time it will reduce any social contact that you have with that individual. In a pack of wolves, it is important that everyone works together for the survival of the pack, and this requires that some sort of social bond be established. A dog that runs away eliminates any chance of social contact. So what is the animal who is fearful of a more dominant wolf to do?

The answer lies in communication. By admitting the superiority or higher rank of the other individual with some sign or gesture, any chance of conflict is avoided. A dog who gives a submissive signal accepts the dominance of the other dog. If that dog then accepts the communication, approaches, and perhaps shows some sign of greeting, not only has a fight been circumvented but there is a chance for some bonding to occur. In this situation, there might be a more restrained version of the greeting dance that we saw between two equal-ranked dogs. The dance is more restricted because the submissive dog stands still. Only the dominant dog moves around to eventually sniff the hindquarters of the lower-ranking dog. Dominant dogs sniff while lower-ranking dogs simply stand and wait. Through this ritual, the submissive dog develops a feeling of trust for the higher-ranking animal, since his position is recognized. He has now learned that if he accepts his lower rank, he can remain in the pack and still be safe from attack.

The presence of a ritualized signal such as the lowered body is not a sign of physical fear, but rather a means of avoiding a fearful situation. A peasant bows before a king to show respect and acknowledge his rank. He knows that by following this ritual, he avoids any danger to himself and may indeed receive benefits, such as protection. The same is true among dogs. This position is the dog's equivalent of bowing before a high-ranking superior.

Taken by itself, this behavior is designed to be pacifying; however, it is often only one of several actively submissive behaviors. For instance, the dog may lick the air at the same time, or provide other pacifying signals as well.

Muzzle-nudge: Lowered body position is often accompanied by a puppylike behavior that we can call the *muzzle-nudge*. This occurs when a submissive dog approaches a more dominant dog and gently pushes against that dog's muzzle with his own nose. This signal, along with a lowered body, signals that the lower-status dog is accepting the higher rank of the other dog.

The evolution of this sign probably comes from the interactions between puppies and their mother that we discussed earlier. Puppies nudge for food. When they are young enough, they nudge their mother's teats to get the milk flow going. Later, they will nudge and lick their mother's face, or the face of another dominant adult, to get them to regurgitate a bit of food. This puppy behavior has now become ritualized into a communication signal, which says: "I know you mean me no harm and will take care of me." We know that this nudge is simply meant to communicate, since submissive dogs will often make the nudging motion without actually making physical contact with the other dog. They nudge the air in the direction of the other animal in much the same way that we might kiss the air while looking in the direction of a loved one simply to show some affection. Dogs often use this muzzle-nudge when they are interacting with people. They will frequently nudge their mistress's hand or leg when they want something, such as food or to go out for a walk. If there is a well-established ranking in the family, they will also often nudge simply to get a bit of attention, and perhaps to be stroked or petted.

Dog sits when approached by another, allowing itself to be sniffed: While the lowered body position is a sign of submission, it is not the only way in which rank differences can be displayed. Take the situation where two dogs meet and both are fairly confident and dominant, although both recognize that one is more powerful. The dog that feels a bit outranked, but is usually dominant over other dogs, may find it difficult to display the full lowered body signal, since that would suggest that the difference in rank is greater than it actually is. Instead, the lower-ranking animal simply sits. By doing this, he eliminates all of the signals associated with threat and challenge, since these require a standing and moving animal. By allowing the other dog to sniff and approach, he accepts the other dog's dominance, but also signals that theirs is not a "royal versus peasant" relationship with a large social gap between. Among human beings, this would be the equivalent of a prince arriving before a king. He may well simply lower his head and eyes for a moment, to acknowledge the king's position, rather than giving the full bow that other members of the kingdom are expected to display.

Knowledge of this signal can avoid confrontations when your dog is out walking with you on leash. If your dog is approached by a dog that seems to be showing hostile intent, you can simply command your dog to "Sit!" If your dog responds, the likelihood of any conflict is usually eliminated. From the point of view of the approaching animal, your dog has acknowledged his social dominance, so there is no need to demonstrate it physically. At the same time, your dog is likely to obey your command without hesitation, since you have not asked it to demonstrate any great weakness in front of this stranger.

Dog rolls on side or exposes underbelly and completely breaks off eye contact: If the lowered body position is the equivalent of a human bowing, then this position is the equivalent of a human groveling. This is the most extreme form of pacifying or submissive signal that the dog can give, since the dog gives up any chance of aggressive action in this position. It is a sign of true social fear, and of considerable difference in social rank. If body postures had a sound, this one would be a whimper, which says to the dominant dog, "I'm just a lowly beast that accepts your full authority." In this powerless position it goes on to say, "To show that I'm no threat, you can do with me what you want."

If the dog really wants to emphasize its degree of social fear and its recognition of a vast difference in rank, it may also release a few droplets of urine. The combination of lying on the ground to make it appear as small as possible plus these drops of urine reminds the dominant dog of puppy behaviors. Puppies need to be cleaned of urine and feces when they are small, and the mother usually simply rolls them on their backs to do this. In effect, then, this groveling dog is saying, "I'm no more of a challenge to you than a helpless puppy is."

When this very passive signal is given, most dominant dogs will sniff the hindquarters of the dog lying on the ground. It is only when the dominant dog turns away or looks away that the submissive dog will begin to move. At this point, it may now assume the submissive body position of Figure 12-3 and attempt to establish some form of social interaction.

We often see parts of this display (without the full break in eye contact or the drops of urine) in less emotionally charged situations. Many dogs adopt this posture in a fairly relaxed and contented manner when they are around their pack leader. The dominant dog may then nuzzle its throat, belly, or genitals, or lick its face, as a sign of acceptance. Dogs will sometimes do this for humans. When your dog rolls on its back, you may think that it is requesting a belly rub. Actually, it is a sign that you have been accepted as the all-powerful leader of the pack (the belly rub is a bonus).

There are other ritualized body positions and touch patterns used by dogs to show dominance. The simplest one of these is *standing over another dog that is lying down.* This is a blatant way of saying, "I'm bigger, taller, and in charge." Adult dogs often stand directly over puppies to make it clear that they are still controlling any interactions with them. There are specific touch patterns that also emphasize this same idea. Dominant dogs, pack leaders, and dogs that have aspirations of becoming a pack leader use a number of ways to say, "I want you to know I'm boss around here." Many are based upon the idea of relative size, since the bigger a dog is, the more likely it is to be dominant. One of the most common involves *the dog placing its head on another's shoulder.* There is a variation of this in which *the dominant dog places its paw on the back of the less dominant dog.* Both of these gestures involve placing a part of their body *over* the body of the other dog. Obviously, a really big dog would touch another dog in this way simply because the smaller dog would be physically lower. However, this gesture has been ritualized to mean that the dominant dog considers the less dominant dog to be physically smaller (even if he is not) and will treat him accordingly.

If a wolf or other wild dog is recognized as leader of the pack, then all other animals in the group will give way to him when he approaches. If he wishes to be in a specific place, he moves forward toward that place and any animal in his path will move aside at his approach. A dog who feels that he is dominant will act in a similar way and will sometimes enforce that clearing of the path. The most active way that dogs do this is with the *shoulder bump.* In dogs, this will often be seen when one dog quickly runs up, approaching the other dog's flank, and then gives him a hard knock with its shoulder. Usually, if the dog doing this action is larger or has a good deal of momentum, the other dog will be driven a few steps to the side, in effect clearing the path for the active dog. In this scenario, the first dog has effectively said, "I am dominant over you and you will give way to me when I come near." He doesn't wait for the other dog to respond but enforces compliance while asserting his dominance at the same time. Obviously, this is a forceful and confident statement.

There is a very subtle variation of this kind of behavior, which often goes unnoticed by humans. This is *leaning.* Leaning is really nothing more than a very passive and quiet version of the shoulder bump. A dog that wishes to express its dominance will move next to another dog and then lean its weight against it. If the other dog now accommodates him, by moving slightly away, then the message that the leaning dog is dominant has been accepted. This body signal has forced the second dog to move and to give up his position, even if that movement is only a matter of inches. Remember, this is communication, not conflict, and the messages sent and received are symbolic. In the same way that a person might momentarily bow his head an inch or two when brought into the presence of a higher authority, royalty, or an eminent clergyman, that subtle motion establishes the relative status of the individuals. No shouting is needed and exaggerated movements are not necessary. One simply has to know how to read the body language.

When human beings interact with dogs, they should be aware of these signals. The leaning signal just discussed is a common subtle way in which dogs try to establish dominance over people (remember the story of Bluto in the first chapter). It is quite common to see a larger dog lean against his master when they are standing together, and dogs who are allowed to sleep on the bed with their masters will often try this signal. If the person shifts position, he has actually lost status, and the likelihood of further leaning will increase. Over time, if this interaction repeats itself frequently, the dog may try other ways of confirming its dominance, perhaps showing disobedience of commands or even more aggressive signals. A large dog that jumps up and tries to place its paws on its owner's shoulders may well be trying to express the same kind of dominance shown by placing paws on the shoulder of a lower-ranking dog.

The familiar signal where *the dog places its paw on its master's knee* may also have the same dominance meaning. However, one should look at this gesture closely. If it is part of a pattern where the dog paws the air in front of the master, tries to slide its head *under* its master's hand, it is probably part of a simple attempt to get some attention. In these circumstances, this signal means, "Look, I'm here," or, "Pay attention to me," rather than, "I think I can lead better than you."

Dogs who want to avoid confrontation, but who do not want to show any great deal of submission, can also use a set of ritualized signals to show that they will accept the current situation but are not accepting the conclusion that they are low-ranking in the pack. Many of these signals are based upon turning behaviors, nonchalance, or distractions.

The simplest way to express peaceful intentions is when *a dog turns its side toward another animal.* Usually it is the lower-ranking dog who does this, but it often is done quietly and without any overt sign of fear or distress, suggesting that the dog who turns away is accepting the authority of the other, but is still confident and in control. Turning to the side often results in the T-position shown in Figure 12-4. Such an encounter virtually never results in any aggression.

A variation of this behavior is seen when *a dog turns its hindquarters toward another.* This is usually a greeting behavior. It shows a bit less

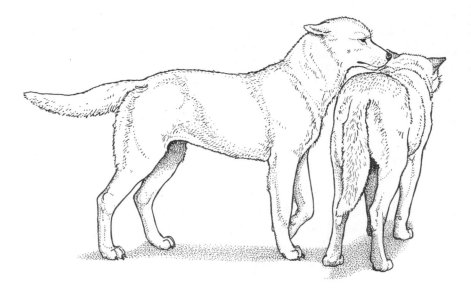

Figure 12-4 The T-position, in which the dog who has turned its side is making a pacifying gesture.

confidence than turning the side, and may result if there is a larger gap between the social ranks of the two individuals.

If a dog is already facing sideways when another dog approaches, and in response to that approach turns to face the new arrival, this is an expression of both dominance and confidence. As with so many aspects of dog language, whether this results in play or conflict is determined by the response of newcomer to this high-ranking gesture.

Some of the Doggish body language that is used to defuse a situation involves pretending to be completely indifferent to the current situation. I have often seen incidents where one dog approaches another in threatening manner only to find himself watching as *the threatened dog sniffs the ground*. For all the world, the dog sniffing the ground seems to be deaf and blind to the approach of the threatening dog. One can rest assured that there is nothing of interest at that spot on the ground, and that the action is simply meant to divert attention away from the actually sniffer. The communication aspect of this behavior is that a dog that seems to be totally preoccupied with sniffing is clearly not preparing some form of aggressive response or challenge. The belligerent dog then has no excuse to continue the threat, since nothing is in dispute.

There are other variations on this distraction and indifference signal. One of these is when *the challenged dog stares intently at the horizon, apparently indifferent* to the other dog's approach. This is simply the visual version of intensely sniffing at the ground. If the threatening dog fails to acknowledge the fact that its target is gazing into the distance, the threatened dog may give a bark or two in the direction it is looking. This invariably deflects the attention of the approaching dog, and almost always ends the threat.

Probably the simplest form of nonchalant indifference in response to a threat is when *a dog responds to provocation by another dog by scratching himself.* This is often the response of a fairly dominant dog to a challenge. I once saw it acted out when a big young Akita decided to confront an older but larger Akita in the park. The young dog began a stiff-legged approach, glaring and staring as he drew near. The older dog simply sat down and began scratching his ear in a bored, indifferent manner. The young dog seemed totally taken aback at this and sat down a few feet from the old dog. The scratching had shown that the bigger dog was not getting ready for a fight, but also that he had no fear of the younger. Now that both were sitting (a moderately submissive signal), it was easy for both to rock to their feet and begin a casual, non-threatening greeting ritual, involving the usual dance pattern with lots of sniffing.

Some of the basic body language of dogs has evolved to convey certain nuances of feeling.

Dog sits with one front paw slightly raised: This is a sign of stress. It combines social fear with a reasonable dose of insecurity and means, "I'm anxious, uneasy, and concerned." In novice obedience competitions, where the dog has to remain in a sit for a minute while its owner stands some 40 feet away on the other side of the ring, you will often see relatively unseasoned and nervous dogs making this gesture. The dogs who do this are also those who are the most likely to lie down or break and run to their master before the time has been completed, thus confirming their anxious insecurity. It is also seen in puppies, where it means not only mild stress but a request to be interpreted as "I need you to do something for me."

It appears that this sign evolved from the signal for submission in which the dog rolls onto its back. If you watch a dog rolling into that position, it begins the roll by first raising one paw and then rolling the front part of the body. This is, then, a fragment of that action, suggesting that the feelings may be fearful but are not strong enough to bring out the fully submissive action pattern.

Not all body language has to do with social rank, dominance, submission, and insecurity. There are many other things that dogs can say with their bodies.

Dog rolls on its back and rubs its shoulders on the ground: This movement is sometimes preceded by "nose rubbing," where the dog pushes its face, and possibly its chest, against the ground in a rubbing motion. It can also be associated with an exaggerated rubbing of the dog's face with its forepaw in a direction that goes from eyes to nose. I like to look at this collection of signs as part of a contentment ceremony.

This little ritual is most often seen after something pleasant has occurred, such as immediately after feeding. Occasionally, although less frequently, the activity occurs while something good is in the process of happening, such as when the dog's master is actually preparing the food. This ritual generally follows, rather than anticipates, a pleasant activity. Thus, our daughter Karen's dog Tessa would give this contentment roll and rub immediately after a romp down by the creek, never before. When she was out visiting our farm, Tessa would often send this happy message after she was let out of the house following a long confinement; however, it didn't come until after she had dashed around the yard to work off her excess energy. Once she had finished her "freedom dance," she would follow it with the contentment roll, and then settle down in her favorite flower bed for a nap.

PLAY

In most animals, playfulness begins to disappear as the individuals become adults. However, human beings have bred dogs to keep many puppylike characteristics, and with this comes the lifelong desire to play. This is important for people, because we humans also retain our childhood curiosity and playfulness through our whole lives. In effect, as perpetually juvenile apes, we have created a playmate for ourselves in a perpetually juvenile wolf.

For young puppies, play is serious business, not just random chaotic behavior. Puppies learn a lot by playing. First, they learn about their own physical capabilities by practicing a variety of maneuvers and contortions. Play also involves many sequences of behavior that are related to escape from danger, self-defense, hunting, and even mating. Most important, puppies learn how to interact with other dogs, and acquire their first lessons in the Doggish language. Their play fighting teaches them about dominance, and they learn to recognize their social position. They also learn what works to influence others—how to get what they want and how to prevent what they don't want.

Play teaches puppies that overt physical aggression is virtually always unacceptable in the social life of their pack. When they first play-bite other puppies, they quickly learn that if they don't use a soft or inhibited bite, bad things happen. For instance, by clamping their sharp little teeth on a littermate's ear, they hear a yelp, and then their friend breaks off the game and they may be chastised by their mother. To continue the social interaction and the play, they learn that actual aggression doesn't work.

Since play involves chasing, biting, jumping, pushing, wrestling, growling, and mock fighting, it is important to signal that these actions are all in fun and not to be taken seriously. For this reason, dogs have worked out a set of play signals.

Dog crouches with front legs extended, rear body and tail up, facing its playmate directly: This is the classic play-bow and the most common signal suggesting, "Let's play!" It is used as an invitation for a romp, and usually is followed by a sudden dash away, or a charge toward the playmate. Chasing and fighting then constitute the major part of the game.

This play-bow is not just an invitation. It actually is a sort of a punctuation mark, which is used during the play period to remind everyone that what is being done is only play. Thus, before a dog charges in the direction of another, in a form of mock attack, it may give the play-bow. If a dog accidentally bumps another dog too hard, or knocks him down, he will usually immediately reassume the play-bow position to reassure the hurt individual that it was just in fun and not meant to be aggressive. Sometimes the play-bow is only part of the invitation. Some dogs will dash madly around when released into an open space. They will prance, leap, jump, zigzag, tuck their tail under their rump and run in wild circles. Interspersed among these exaggerated movements will occasionally be a quick play-bow, which is immediately abandoned as the play-seeking dog dashes off again to twirl and prance. This zany behavior actually has its basis in a hunting strategy adopted by wolves and foxes. By "dancing" in an unpredictable manner, they capture the attention of animals that might be prey. As these misguided creatures approach the canines to try to discern the meaning of their apparent madness, they can be taken from ambush, or may be lured close enough to be pounced upon.

In North America, over the last century, this strategy was used by hunters to attract ducks. They encouraged their dogs (originally poodles) to caper around in a foolish and playful manner in an open area. When wild ducks caught sight of this activity, they became fascinated, and flew closer to investigate this absurd scene. Such would be their undoing when they were drawn close enough to be within shotgun range. To catch ducks in this way was called "tolling," from the French word *tollen,* which means "to entice." Thus, we toll the church bell to entice or draw people to religious services or to rally the community in a time of trouble. Later on, Canadians developed a dog specifically for this kind of hunting. The Nova Scotia Duck Tolling Retriever (called simply a toller by its fanciers) will not only dash around on land to attract ducks, but will also swim around erratically in water for the same purpose. Once the duck is shot down, however, it does the work of a regular retriever.

The play-bow is sometimes not enough to draw some timid young dogs into play with adults. Older dogs seem to find this quite frustrating and will go to a fair bit of trouble to tempt the puppies to play. This sometimes involves using a signal that has another meaning as a sort of "reassurance signal." The most common behavior involves a dominant dog approaching the young one and rolling over on its own back in what looks like a full passive submission display. By adopting this sign of lower rank, they seem to be saying to the younger dogs, "You can be leader for a while if you play with me." Perhaps feeling a bit self-important at having a larger, older dog act submissively, the young dog approaches. Once he is close, the older dog rolls into a play-bow, and the romp can now begin.

Dogs don't play many different varieties of games, but the ones they do play they play with great enthusiasm. "Keep away" is probably the most popular, and it involves picking up an object and dashing away in hopes that it will be chased. Sometimes, the dog with the object will approach to within a few yards of its playmate and drop the object, just to attract the other to make a run to get it. The moment he does so, the object is immediately grabbed again, and chase begins anew. Often the object is dispensed with, and the game becomes "Tag," with one dog chasing the other. Who chases whom will often change, and when a dog is caught, the game changes to a perennial favorite, "Wrestling," with enough noise and growling to convince people who don't know much about dogs that someone is about to get killed. Another game is "Charge!" where one dog runs straight at the other, only veering away inches from its target. This looks very threatening, but when it works, it will rapidly turn into "Tag," with the target dog chasing the charger.

To watch dogs run in play is to appreciate grace and joy. It is also a key to understanding something about their psychology: Running is to dogs what dancing is to people. It is their way to get into the rhythm of the universe.

13

The Point of the Matter

In chapter 4, when we looked at some reports about dogs that supposedly had very good linguistic capabilities and very large receptive vocabularies, we found that, in some cases the dogs were really responding not so much to the words as to the body language of the human beings who were uttering those words. Most of these situations involved the dogs reading what the person wanted them to do, or where the person wanted them to go, by some form of subtle body turn or gaze in a particular direction. The first time we encountered this phenomenon, we viewed it as a sort of contaminant, which was distorting our ability to determine how many spoken words a dog could understand. However, it is possible to rethink this conclusion in light of what we now know. Since we have seen how well dogs read body language, perhaps we ought to reinterpret their pointing behavior as part of the dog language of body signals. A significant reason for doing this is because dogs not only interpret the body language of others in terms of pointing, but also actually produce pointing gestures with the obvious intent to communicate. In other words, pointing is both a receptive and a productive part of their language.

When I speak about the specialized aspect of body language that we term *pointing,* I don't mean the classic pointing of a hunting dog, such as a pointer or a setter, who orients its head and body, then freezes, to hold a position pointing to where the bird is. The kind of pointing I am talking about is much like the pointing that people use to inform others about the world. It may be useful to look first at the equivalent behavior in humans. The average person may not think of pointing as language, but when scientists look at human language development, they find that pointing has many similarities to language.

Some psychologists feel that the first word that a child utters is not a word at all, nor even a sound. It is a *gesture*—namely, finger-pointing. When you point at an object, you are not saying anything about your finger at all. Rather, you are indicating a specific object at a specific location in space. Thus, if you point at a shiny piece of jewelry on a table, in effect you are saying, "Look at that piece of jewelry." An alternative is that you might be saying, "I want . . ." or, "I like . . ." or, "I'm interested in that piece of jewelry." But you are certainly not saying, "Look at my finger."

Human children are not born with an ability to point meaningfully at objects. If you show a nine-month-old child an interesting toy or a cookie that is beyond its reach, it might first stretch its arm out with all fingers extended toward the object. The child will look directly at the thing it wants, and when it can't reach it, may display frustrated behavior—banging on the chair or table, making noises, and so forth.

Sometime around ten or eleven months in girls or around thirteen to fifteen months in boys, there is a sudden change. Now the child no longer gestures with fingers spread but begins pointing. That this is an act of communication is exemplified by the fact that if there is no one in the room with the child, he or she will not point. Furthermore, the child will tend to look at its parent or at another adult before pointing, and often even while pointing. At the same time, the child may try to create a wordlike sound. These sounds the child is making may be an attempt to put a name on the object, or may just be a means of getting a nearby parent to look toward him, so that the adult will also become aware of the pointing gesture. Notice what is going on here. The child is designating an object in a particular location and trying to communicate, "I want that thing which is over there." The finger point serves the same function as the label of the object. Thus, if the child points at a cookie and we bring her the cookie, it has the same effect as if she had uttered the word "cookie." In this way, we may look at the pointing gesture as the first word or some kind of protoword.

I had a conversation with a psychologist who studies the development of language in children, and she informed me that it was pointing behavior which convinced her that dogs can never have anything like language.

"When I point a finger to show my dog where something is, even if it is something she really wants, like a treat, what do I get? My dog looks at my hand. If I hold the point in place, she'll come right up to my hand and start butting my finger with her nose. I can point to things a dozen times. She simply gets more and more frustrated, but still keeps coming back to my hand every time. The idea that my pointing finger designates 'that treat over there' just never occurs to her."

There are two problems with this analysis. The first has to do with the fact that we always assume that dogs and other animals must act exactly the same way, and use the same tools, that humans do to accomplish similar results. This assumption is clearly impossible here. Dogs don't use their paws the way people do. They don't do much in the way of manipulation, and they certainly don't use their paws to gesture. When a dog points, it is not with a paw, but with its head and body. When my dog Odin wants to go outside, he looks at me and then orients his head and his body toward the door. This is the equivalent of finger-pointing in a human. If I don't respond, he will look at me, give a tentative bark, and again look at the door and orient his body in that general direction.

While a dog may learn to respond to human finger-pointing, its natural instinct is to look for head and body turns. I demonstrated this to my own satisfaction by using Odin. There is an exercise in obedience competition called the "Directed Jump." As part of the competition for the Utility Dog degree, on command, the dog is required to run forward and at a distance of about 40 feet until it is instructed to turn, sit, and face its handler. On either side of the ring is a jump: the handler must indicate whether the dog is to take the left or the right jump on its way back to the handler. This is usually done with a broad pointing motion in the direction of the desired jump while at the same time the handler gives a verbal command, such as "Jump."

I was just beginning to train Odin to jump on command, and at the time when I began my little experiment, I had never had more than a single jump in the ring. For my experiment, I first placed two identical jumps midway in the ring, separated by about 10 feet. Placing Odin at one end of the ring and myself at the other, I deliberately turned both my head and my body so that I was oriented toward the jump on the right, at the same time calling out: "Odin, jump!" Without hesitation, the big black dog shot across the ring and over the jump that I was oriented toward, and finished up in front of me. When I oriented toward the left jump in this way, he was equally accurate. Obviously, using both head and body orientation tells the dog quite clearly where we want him to do something.

Next, I placed him into position at the other side of the ring and gave the same jump command. However, this time I kept my body and head straight, and only turned my eyes toward the jump on one side. Odin rose quite slowly, looking back and forth between the two jumps and then at me, obviously searching for a clue, and clearly not getting it from my eyes. He was moving toward me, down the middle of the space between the jumps, and he appeared puzzled and upset. I quickly terminated the test, calling him directly in to me, rather than leaving him out there in confusion.

The next time, I returned to giving a quick orientation of both head and body to indicate the next jump, and he demonstrated to me that he was still paying attention by jumping the one that I had turned to. Obviously, without any training he could pick up directions indicated by my body, but not my eyes. To see what else he could respond to, I moved on to an intermediate level. Instead of orienting both my head and body, I kept my body straight and only turned my head toward the side of the jump I wanted him to take. Again he rose quite hesitantly, looking at me for information. I kept my head oriented toward the jump, and as he drew closer he became more confident, ultimately veering to the side I wanted just a foot or two in front of the barrier.

Clearly, he had used my head orientation to read what I wanted, but it wasn't as good a signal as my head and body together. Yet wolves can apparently go in the direction signaled by the pack leader's head turn, even at a great distance. I had seen and analyzed films of wolves doing this. Why, then, was my head turn so poor a signal? It finally dawned on me that wolves have a long, tapered muzzle that makes it unambiguously clear in which direction their head is pointed. We humans have only a relatively small nose. Although the turn of our head might be clear to a dog if it is close enough to make out the direction that our nose is pointed, at a distance of 40 feet this is less clear. If, however, I had a full muzzle, that would be visible enough at a distance for my dog to be able to determine the appropriate direction to go in.

Taking advantage of the fact that there was no one at home who might worry that I had finally gone over the edge into full-blown insanity, I went into the house and constructed a dog's muzzle. Well, actually, it was a foot-long cone of white paper, to which I could attach some elastic bands so that I could keep it on my head. I darkened the pointed tip of the cone by blackening it with a felt-tipped marker to make it appear more like a canine muzzle with a black nose at the end. When placed over my own nose, it made me look more like some surrealistic bird than a wolf, but I reasoned that the principle was what was most important rather than the aesthetics.

Back out to the field I went, placing Odin at the far end of the ring. Now, with a deliberate turn of my large snouted head, I called: "Odin, jump!" Without the least hesitation this time, he trotted in the direction of the jump I was looking at. Repeating the test for the jump on the other side proved that this was no fluke, that he could, and did, easily read my head position.

After the third jump, I stopped the test. Actually, Odin stopped the test. As I was bending to praise him for taking the jump, I nearly poked him in the eye with my paper beak. He responded by defensively grabbing my paper muzzle in his teeth. He pulled at it for a moment, but couldn't remove it because of the elastic bands that were holding it in place. Somewhat distressed, I shouted, "Odin, drop it!" which is the command that causes my dogs to drop whatever they have in their mouths. Obediently, he released my fake nose, and from the distance that the elastic had been stretched, it came smashing back into my face. The resulting bruising and headache convinced me to call off my experiments for the day.

On the basis of this little study, I pointed out to my psychological colleague that if, instead of pointing toward a treat, if she would move her body so that it was oriented toward the target, lean somewhat forward, and stare intently at the appropriate place, it should have the same effect on her dog that pointing has on humans. She was skeptical, but willing to try, and she invited me to her home to watch. Her springer spaniel, Sally, responded to finger-pointing exactly as my colleague said she did, by watching her hand instead of looking at the treat I had surreptitiously dropped in the middle of the floor. When the psychologist did the "body and head point," however, Sally turned to align herself along the sight-line of her mistress. She proved this by spotting the treat and dispatching it immediately.

The second reason why my colleague's initial comparison was unfair to dogs is because it is likely that some aspects of pointing behavior are learned phenomena. Human parents and their infants interact with a large number of pointing responses. The parent will point at a nearby cat and say, "See the cat?" The parent might point to a visitor and say, "Look, there is Aunt Sylvia." The parent feeding a child might have two types of food in front of it, and point to one and say, "Would you like these carrots?" Then, shifting the direction of the point, continue with, "Or would you like these peas?" Numerous interactions such as these can teach the child the significance of finger-pointing.

Evidence for the learned nature of pointing comes from what are sometimes referred to as "closet children." This label rather starkly designates what seems to be an increasingly widespread phenomenon in Western society. Social workers often euphemistically term children (usually preschool age) who are left alone by their parents without any supervision or other social contact as "abandoned at home." They are often shut in a small room, or occasionally even a closet. The parents often explain that this is "to protect them while I'm away at work," or "to keep them from doing anything dangerous, or making a mess, while I'm away." Such children wait in a form of social and sensory deprivation for the entire day, while their parents are off somewhere else.

Apart from the severe emotional and social damage that it causes, this kind of cruel rearing deprives the child of the environment that is needed to develop language. To learn to speak, there must be someone who serves as a model speaker of the language and who answers your speaking with their own speech. It is not surprising that when such closet children are found, they often have virtually no speech or language skills. In most instances, these children also do not show finger-pointing, despite the fact that they may already be four or five years of age. Instead, they remain at the stage of hoarse yells, with hands extended and fingers spread toward the object they desire. This suggests that pointing behaviors in humans, like other communication behaviors, must be learned. It is proven by the fact that one of the first signs that the child can still learn language, now that it is "out of the closet" and into a normal social and sensory environment, is the appearance of pointing behavior.

If humans must be taught pointing, then why should we expect that dogs would respond to pointing without training? The fact that dogs can be taught to respond to pointing is shown in the directed jumping exercise that I just mentioned. After the dog is trained, all that one needs do is point an arm in the direction of the desired jump for the dog to know which one it should go over. Turning your head and body toward the jump is actually illegal during the competition and could result in disqualification.

In some cases, pointing has advantages over spoken language. It can be done quite surreptitiously. Sounds can be picked up by everybody who is near the speaker, and, for a hunter, like the dog's wild cousins, this means that not only pack members but also the quarry may hear the sound, use it as a warning, and escape. Pointing is silent and avoids this wide broadcasting of the message. Only individuals who have a line of sight that includes the pointer will get the message. Furthermore, if the pointer uses only small, subdued pointing gestures, the message is even less likely to be picked up by individuals for whom it was not intended.

Let's take a simple human example. My wife has little love for the formal types of receptions and cocktail parties that I'm often expected to attend as part of my obligations as a university professor or an author. In most cases she simply doesn't attend, but sometimes, when the event is local or important, she will come along. At those times, it is not unusual for me to glance up and see her communicating to me with a subtle set of covert pointing gestures. For instance, she might point to herself, and then to a chair or a group of people, indicating that that's where she will be found if I am looking for her. More often she will point at her wristwatch, and then at the door, to indicate that she feels it's time to go.

Whole conversations, and the coordination of complex activities, can be achieved through pointing. I recall one such situation which occurred during my combat training in the army. The situation was one of those simulated combats, where one group of soldiers plays the role of defender and the other has to mount an assault against them. The task of my platoon was to gain control of a small hill, which was fortified with a few hidden gun emplacements. My eight-man rifle team (which also included one monitor or referee who would determine who was declared dead) was supposed to clear out any defenders who might be on the eastern slope of the hill, which had the most cover (in the form of bushes, trees, an old damaged stone fence, and some rain gullies).

We silently followed our sergeant, a tough career soldier by the name of Tyner, through the available cover until we were near the base of the hill. Suddenly, he stopped and pointed up the slope. Looking along the line of sight, we could see what looked like a four-man machine-gun emplacement encircled with sandbags. Somehow, we had come quite close to our enemy, and were now less than 100 feet away. Any sound might reveal our location and expose us to machine-gun fire (which would quickly terminate our participation in the exercise).

Sergeant Tyner pointed to each of three men, who then came up to him. He next pointed along the line of the stone fence. His arm traced a straight line away from him, and then the pointing finger swung up and to the side, clearly indicating that these three were to work their way along the fence line, then around the corner where it changed direction and began going up the hill. Next, the sergeant pointed to his watch. He pointed to the minute hand, and then to mark on the clockface which would represent the passage of about ten minutes. He pointed at a rifle, then toward the gun emplacement, then back at his watch. The message—that these three men were supposed to move along the fence, around the corner, and in ten minutes were to begin firing their rifles at the enemy—could not have been clearer. He pointed again to the fence, and the three soldiers began to move quietly toward their position.

Next, Tyner pointed to each of two men who were equipped with rifle grenade launchers. He pointed to the launchers, and then down to the ground. Both men knelt down and began to fix the grenades to their weapons. The sergeant pointed again at the two men, then at his own eyes, then to the enemy bunker, indicating, "Stay here. Watch for my signal to fire."

Finally, he pointed to the remaining two of us, gave a quick small swinging point in the direction that we were to move, and we began quietly to follow him. The referee decided to tag along with our group to monitor the action. We kept to the available cover, and proceeded several yards to the side and then some more distance up the hill. Because the sergeant had to coordinate activities, he always kept himself within sight of the two men with the grenade launchers. We finally reached a safe position and waited.

Some minutes passed until, exactly as planned, the three men near the curve in the wall began firing their rifles. The men in the gun emplacement immediately turned their weapon to that side, so that their backs were toward us, and directed their attention to where they thought the attack was coming from. After a long minute had passed with this exchange of rifle and machine-gun fire, Sergeant Tyner pointed at the distant men with the rifle grenades. Two cracking sounds came from their positions, and before we could even hear the grenades land, he pointed to us and the emplacement, and we began a mad charge across the remaining few yards. We arrived before the confusion of the machine gunners had cleared. The monitor declared the emplacement captured; all of its occupants were ruled to be casualties.

The important point is that this whole coordinated sequence of behaviors was organized and controlled only by pointing. I do not recollect one word being spoken from the moment that the sergeant first spotted the enemy position.

Roughly thirty-five years after this incident, I had the opportunity to see a similar scenario played out, only in this case the actors were all canines. There are several research projects in North America that have been set up to study the behavior of wolves. Several libraries have collected films and videotapes of wolf behavior as observed by these projects. As I was going through one such set of films, I found a sequence which showed wolves hunting. The similarity between that incident and my own military training experience was remarkable.

The species being studied was the Grey Wolf (Canis lupus), which many scientists believe is the wild canine that is genetically closest to our current domestic dog. Just because they are called grey wolves does not mean that they are necessarily gray in color, and this particular pack of six animals varied from a creamy whitish to a sandy yellow-gray. It was midsummer, there was lots of foliage, and the wolves were quietly resting beside a small clump of scrubby trees. There were four adults, two males and two females, and two juveniles. The pack leader (usually called the Alpha male) was a very large animal, weighing around 175 pounds and standing around 30 inches at the shoulder; the lower-ranking male was about 25 pounds lighter. The two juveniles belonged to a mating between the Alpha female and the Alpha male. The Alpha female was also large, at least for a Grey Wolf, weighing around 135 pounds, and she seemed to be the first one of the pack to catch the scent of some nearby deer. She rose, sniffing the air. She took a step forward, brushing lightly against the dominant male. She looked directly into his face, then toward the location of the scent, in a classical canine pointing gesture.

At this point, the pack leader took over the coordination of activities. He rose quickly and looked along the line of sight indicated by his mate. He assumed a position next to her right side, but about a step in front. He glanced back to the lower-ranking male and then pointed his head in the direction of the deer. The other male moved forward to take up a position to the right of the leader. Meanwhile, the other female and the juveniles observed this behavior and quietly moved to positions to the left of the dominant female. All were oriented into the wind, toward the deer, and the lower-ranking female and the juveniles continually checked the direction that leader was looking and tried to orient themselves accordingly. The whole pack seemed to be crowding together, and at first glance it looked as if they were touching noses, as is typical in canine greeting ceremonies. However, careful observation showed that they were really just trying to get close to the leader so as to orient their heads and bodies exactly along the line that he was pointing with his head. The head-pointing served the same function as Sergeant Tyner's finger-pointing: the position of the quarry was being indicated to the group.

Now the wolves silently moved along the pack leader's line of sight. As they approached a clearing, I could see that they were pointing to two deer which were browsing in the open. One was an adult female and the other a yearling. The pack leader looked directly at the other male, then looked down at a place on the ground, just to the right, and level with his own shoulder. The lower-ranking male immediately moved to that spot. The movement was as exact as if the leader had drawn a chalk-mark on the ground.

Having positioned the male, the leader next looked directly at the dominant female, made eye contact, then shifted his gaze to a point to the far left, near the opposite side of the clearing, leaning his body forward in the direction of his gaze. The Alpha female looked at the two juveniles and the other female, and began moving in that direction. These four wolves quietly picked their way along the edge of the clearing, hiding behind the leafy bushes. Every few yards, the dominant female would stop and look back at the pack leader. He was carefully eyeing the grazing deer, but when he noticed her looking back at him, he would immediately turn his head to look directly at a point near the far edge of the clearing. She, in turn, would follow his gaze, and move in that direction. In my mind's eye, I could see the three riflemen from our squad following Sergeant Tyner's pointed direction, to take up positions along the stone wall.

When the Alpha female and her group had reached the designated point, she again looked back at the leader. Now he looked at her and then directly down at the ground in front of him. The distant female looked down at the ground in front of her, at which point she and the other three animals in her group quietly lowered themselves into a crouching ambush position.

The pack leader looked directly at the male to his right, then sharply back at the deer. At that moment, both males sprang instantly into the clearing as if shot from a cannon. At full speed they raced toward the browsing pair. As soon as the deer saw the two wolves, they turned to flee, heading toward the far side of the clearing. At that moment, the Alpha female sprang the trap, leaping into the clearing accompanied by her three companions. The deer could not react quickly enough, and the dominant female leaped on the back of the yearling. She was quickly joined by the other female biting at its hindquarters. The action of the two females slowed the yearling, causing it to veer from the shortest path to safety, and a moment later both males converged upon it and quickly killed it. The two juveniles began to give chase to the surviving deer, but when they saw that the rest of the pack was not joining them, they quickly returned to take their share of the day's hunt.

The similarities between this incident and my own training experience were a bit eerie. The entire strategy—the initial attack to draw the attention of the quarry, followed by a flanking maneuver—was almost identical. Even more startling was the fact that the entire action had been communicated and coordinated without a single sound. In each case, every message was a gesture involving a finger or arm movement in the case of the human attack, or a head or body movement in the case of the canine attack. Nearly every one of those gestures was in the form of pointing. For both the humans and the wolves, these pointing gestures had specific meanings, and were clearly designed to communicate and coordinate a group action. Both humans and wolves used the pointing gestures to mean, "Look, that is our objective over there," "Go to that place that I am indicating to you," and, "Assume a position [kneeling or lying down or at a specific location], "Wait here," and "Attack now." The wolves, of course, had nothing in their conversation that said anything about equipment (since they had no rifles or grenade launchers) nor did their messages address the issue of time (since they had no watches, and certainly no idea of what was meant by "ten minutes"). Other than those technological considerations, the "conversations" were remarkably similar.

It should be clear, then, that although pointing behavior may be a very primitive from of communication for young humans, it can evolve into a complex form of gestural language. Dogs and their wild cousins have this same linguistic ability, and their pointing behaviors have evolved into a highly complex form of communication. If pointing is taken as evidence that children are trying to communicate by designating objects, and are putting some temporary label on them with their gesture, then we must also conclude that dogs are capable of some form of basic object designation as well. They can point (with body and head) and interpret the pointing of another animal. Responding to body and head pointing seems to come naturally; however, the interpretation of human finger-pointing seems to require some specific training.

14

Sex Talk

Adele was around forty-five years old, and as she came up to me, she looked very uncomfortable.

"Since you're a psychologist, I was hoping you could give me some advice."

I immediately anticipated that she, or a member of her human family, had some problem. Most people only seem to mention that I am a psychologist when they are dealing with human problems. The rest of the time I'm usually just the "Dog Man."

"It's my Samuel. I'm afraid that he's gay, and I'm hoping you can tell me what to do about it."

This was not the first time I had encountered this situation, and I was preparing to launch into my standard opening comments about how young people sometimes experiment with alternate sexual practices before settling into the usual heterosexual behavior, and even if her son was adopting a homosexual lifestyle, contemporary society is much more accepting of these behaviors, and many practicing homosexuals lead quite happy, productive lives . . . I never did get that far because it suddenly dawned on me that Adele didn't have a son, and her husband's name was Roger. She did, however, have a boxer that she usually called Sammy.

I asked carefully, "Just exactly what is Sammy doing?"

"Well, a couple of days ago when I was in the park he tried to . . . you know." She sighed and started again. "He tried to do it to my friend Nancy's golden retriever, Benji. We were both embarrassed and I knocked him off Benji's back and everything seemed to calm down. Yesterday, I was in the park again with him, and he grabbed the back of a Labrador retriever that I'd never seen before and began trying to . . . you know, he began to try to do that sex thing to him. I thought he'd just taken a fancy to some girl dog who wasn't really cooperating and seemed to be rejecting him. The next thing I knew, his owner was rushing at both dogs. She was actually screaming at me, 'My Walter is a normal boy dog. Get your filthy homosexual animal off of him! Somebody help me!'

"She was shouting and then she started to hit Sammy with her leash. I mean, people were looking at all of this commotion—they were actually staring at us! It was really embarrassing. I got the dogs apart and took Sammy away. I didn't take him to the park today, and I don't know if I can ever take him back there again if he's going to act like that."

The significance of Sammy's mounting these other dogs had been completely misinterpreted by Adele and the retriever's owner. Both of these women, and others in their lives, were really taking their own attitudes toward sexual behaviors and applying them to dogs. For many people, all they know about sex in dogs is that it occurs in a position known to humans as "doggy-style." There is much more to canine sex than this, and much more significance to mounting behavior than sex.

There are a few facts that everyone should know about sex among dogs. First of all, there is a clear inequality between males and females. In humans and some (but not all) apes, both the male and the female are sexually active all through the year. In the vast majority of other animals, both the male and the female have a "season," which is a brief period of time when both are ready for intense sexual activity. With dogs, however, males follow the first pattern, which means they are ready for sexual activity throughout the year, while the females have only two relatively brief periods of "heat" when they are interested and ready for sexual activity.

One might be tempted by these differences to suggest that male dogs spend most of their year in a state of sexual frustration, surrounded by usually unwilling females. This is not the case. While males may be interested in sex at any time, they actually only become sexually aroused in the presence of a female in heat, or at least in the presence of the scent that she gives off. It is during heat (technically called the *estrus* period) that the female's ovaries begin to produce the various sex hormones needed to make her fertile, as well as the particular scent that makes her attractive to male dogs. The term *estrus* comes from the Latin word for "frenzy," since these same hormones also make her much more active, and sometimes more dominant and aggressive.

The estrus period lasts about twenty-one days and is divided into three stages. The first stage, *pro-estrus,* typically lasts around nine days. During this stage, the female begins to show a lot of restlessness. She tends to wander more than usual. She drinks more than usual and urinates a great deal as she wanders. It is the fragrance of this urine that attracts the males. They sniff it and then raise their heads and seem to stare into the distance as if they were contemplating some great philosophical mystery. Males can detect the female's odor from a great distance, and it is not unusual to see a bitch in heat being avidly pursued by a large group of prospective suitors who gather hopefully around her home.

As the pro-estrus period nears its end, the vaginal discharges begin to be dark and bloody. It is at this point that many people incorrectly say the female in menstruating. Menstruation in humans occurs well *after* ovulation. It marks the end of each period of fertility and represents the breakdown of tissues not needed to support a fetus when the female is not pregnant. In dogs, the bleeding occurs *before* ovulation and represents changes occurring in the walls of the vagina to ready her for ovulation.

During this period of time, at least from the viewpoint of the hopeful male dogs that are attracted to her, the female earns the street meaning of her technical name, "bitch." She has been spraying that inviting sexual perfume in her urine, and it has been wafting in the air from her vaginal discharges and attracting every male dog in the vicinity. Nonetheless, she rejects every amorous advance from males. She may growl at the hopeful lover, threaten him, chase him, even snap at or bite him. A less aggressive female often simply runs away, or spins around when he tries to mount her, so that the panting Romeo ends up facing a threatening face rather than an inviting rear end. A less complicated strategy she might adopt is simply to sit down, thus cutting off any access to her rear.

This is not some kind of teasing game that the female is playing. She has not yet ovulated, and she will not do so until around the second day of the estrus cycle proper, when the discharges become clearer and more watery, indicating that the vagina is ready for mating. Once the eggs are released, they will need around seventy-two hours to mature before they can actually be fertilized by the male's sperm. Her fertile period might last only a few days, so it is vital that she has attracted enough males around her so that she can select her mate at the proper time.

Courtship behaviors and signals have a lot of features which are very similar to play behaviors and signals, with a few special invitational gestures. For the most part, the female is in complete control. This is to be expected, since it is the female that must invest so much energy in the process of conception, growth of the puppies in the uterus, birthing, and postnatal care. In the wild, this means that an active selection process is taking place, in which some candidates for fatherhood are forcibly rejected while others are vigorously pursued. Evolution has instilled a certain degree of programming, which encourages the bitch to choose a dominant, strong dog that will pass along fit genes.

At this point it is important to point out the difference between the sexual behaviors of modern dogs and wild canines. During the process of domestication, we profoundly changed the nature of dog reproduction. To be exact, we created an animal that is much more fertile. Except for Basenjis, domestic dogs usually come into season twice a year as opposed to once a year for wild canines. Our domestic dog is also much more promiscuous than its wild cousins. This was a deliberate and conscious effort on our part, and is a necessary aspect of our desire to create dogs with specific characteristics.

Each breed of dogs comes about through selective breeding. This means that we must be able to take a dog which has certain characteristics (such as a particular coat color, body shape, or behavioral ability such as retrieving or herding) and mate it with another dog with similar or other special characteristics that we desire. Obviously, having a dog that comes into season more frequently gives us more opportunities to try to mix and match the genes from various dogs that might produce exactly the dog we want. However, it also ought to be equally obvious that for such controlled breedings to be successful, the prospective canine parents must be willing to accept each other as sexual partners. If domestic dogs were highly selective and rejected mates that were chosen for them, this would stand in the way of the creation and maintenance of our modern dog breeds. For this reason, promiscuity in dogs is actually a desirable characteristic.

This is not the case in the wild. Among wild canines, indiscriminate breeding would be a disaster, since it would put so much pressure on local food sources. A wolf pack will usually have only one litter, averaging four to six pups, and this will usually result from a mating of the Alpha wolf and the Alpha female. If times are bad and food is scarce, even that one litter might be skipped.

The courting behaviors among wild dogs, and less frequently among domestic dogs, can go on for hours. Sometimes, in the wild, the courtship might even be temporarily suspended and then continue the next day. The female will usually start the mating dance by running toward the male, then dashing away, then running back at him, followed by a quick retreat again. Most males find this behavior irresistible, but in the unlikely event that the male is playing "hard to get," the female may begin to frolic around him, sometimes actually striking him with her paws. If that doesn't work, some females will actually attempt to mount the male, as if she were reminding him what this whole game was about. In the end, there is usually an extended period of chasing, dodging each other, scampering this way and that. These activities are often interrupted by play-bows, and some periods where the two dogs will rear up while facing each other, sometimes placing their paws on the other's chest or shoulders and pushing away, as if wrestling.

After this period of frisky cavorting, the potential pair comes together and they begin to explore one another's bodies. This usually starts with a few bouts of nose-to-nose sniffing, which may then be followed by some ear licking. Finally, the pair moves to the sexual parts, beginning with some extended sniffing of each other's rumps. Now the female calls the shots. If she is interested, she tells the male that she is ready by presenting her rear and moving her tail to one side. At this gesture, the male will generally check out her intention one more time by coming up to her side and resting his chin across her back. This is the crucial moment. If she stands steady and stiff and makes no sign of moving away, the male then swings around and mounts her. He leans over her back, grabs her hindquarters with his front paws, and begins thrusting. This is the position assumed by dogs when they are actually copulating.

In the wild, when the females in pack are in season, there is a lot of courtship activity; but since wild canines are much more selective, there is a lot less actual mating. One researcher studied wolf-mating behaviors in a single pack during the spring estrus season and counted 1,296 courtships in about a month. These ended up with only thirty-one completed matings, which means that only 2.4 percent of all of these courtships ended in a consummated sexual relationship.

It appears that our domestication of dogs has not changed much in the sequence of courtship behaviors, since dogs, wolves, coyotes, jackals, dingos, wild dogs, and even foxes seem to use the same courtship dance. In domestic dogs, however, the length of the courtship period has been greatly reduced. Even more importantly, there has been a tremendous increase in the likelihood that each courtship will end in an actually completed mating. When pedigreed matings are arranged, there are still some occasions where there are refusals, but these are now sufficiently rare that when they occur they tend to cause concerned discussions among the breeders involved.

In true sexual behaviors, the mounting only occurs at the very end of the mating dance, and then only if the female has quite explicitly accepted her male suitor. Compare this to the mounting behavior that occurs between two male dogs. It is usually preceded by careful sniffing, followed by stiff-legged movements with the tail and ears in quite an erect position. This is not the play behavior that starts the mating sequence. Obviously, then, when a male dog mounts another male dog, the implications and the message have little to do with sex.

The fact that mounting behavior can be relatively independent of sexual intentions can be seen by watching very young puppies. Well before they have reached puberty (which comes at about six to eight months of age), they are already showing this kind of activity. Mounting in puppies appears shortly after they begin walking and is common when they start playing with each other. It is a socially significant behavior, not a sexual one. For young puppies, mounting is one of the earliest opportunities for learning about their physical abilities and their social potential. It basically represents an expression of dominance. The stronger, more authoritative puppy will mount its more submissive brothers and sisters simply to display leadership and dominance. These behaviors will then carry on into adulthood, the significance being power and control, not sex.

This mounting behavior is used as a signal for dominance, and since it is unrelated to reproduction, its social significance applies to both males and females. As a display that serves to challenge or to assert social dominance by one dog over another, this behavior can occur between individuals of the same or the opposite sex. A male mounting another male is not displaying homosexual tendencies, but is simply saying, "I'm boss around here." Females may use mounting as a statement of social position as well. Females can be dominant over other females and even over male dogs, and can display this by assuming a mounting position. This is not an issue of sexual confusion, since the dynamic structure of dog society is not a question of gender alone. Status in the canine world depends more upon size and physical ability, combined with certain characteristics associated with temperament, motivation, and drive.

There are three different hierarchies in the social structure of dogs. There is the overall rank in the pack, which starts with the leader at the top and moves down to the ultimate underdog. There is a lead or Alpha male and an Alpha female, and one of these will be overall pack leader. There is also a ranking among the rest of the males, and another among females. Mounting behavior may occur to assert any one of these rank orders, which means you may see males on males, females on females, males on females, or vica versa. None of these behaviors represents any form of sexual advance or invitation. Instead, they should be viewed as a very clear signal of serious social ambitions by the mounting dog. Some of the dominance expressions that we saw earlier, including the dominant dog placing its head or paws over another dog's neck or shoulders, may well be simply subtle components of activities related to mounting. The dominant or "top dog" is literally the dog that is on top.

Since mounting behavior is most commonly an attempt to claim a higher social status than that of another animal, it should not be surprising to find that the belief that you can stop your dog from mounting by neutering him is just a myth. Neutering will eliminate certain sex-related hormones in the dog, such as testosterone, and the reduction in these male hormones will tone down the dog's aggressive tendencies and also reduce some of its other dominance behaviors. So it may reduce the appearance of mounting behaviors. However, neutering will not change the dog's basic character and personality, which means that in a dominant, leadership-oriented dog, mounting behavior may still occur. What the removal of the sex hormones will do is reduce the intensity with which the dog will pursue his social ambitions. However, the older a dog is when neutered, the less his dominance traits will be curtailed. Castrating a male dog will eliminate his ability to reproduce. Erection will still be possible, but sperm production has been stopped. This means that the dog may still be interested in a female in heat, but any such attempted matings will be literally "fruitless."

Though mounting behavior is not something that many people find acceptable in their dog, in comparison to dogs actually fighting, with a full display of gnashing teeth and slashing attacks, it is really quite controlled and harmless.

A number of scientists have recently begun speculating about the link between male sexual behavior and dominance in humans as well as dogs. Some of this speculation has been triggered by the present rash of press coverage of prominent politicians caught engaging in embarrassing extramarital affairs. These scientists note that politicians are certainly socially dominant individuals, and then openly ponder about the possibility that leadership characteristics might be linked to a biological tendency toward increased sexual behavior. The more dominant the individual, the more likely there will be promiscuity, whether socially accepted (as in the case of cultures where men can have more than one wife) or not. Didn't King Solomon have more than one thousand wives, according to the Bible? In a typically scientific manner, these researchers have suggested that it might be possible to separate sexuality from dominance by using some highly specific drugs that can nullify the effect of certain hormones. In this way, it might be possible to create politicians with strong dominance but no interest in bimbos, White House aides, or night-club entertainers. It should also be possible to create people of the opposite stripe, socially irresponsible and promiscuous, something like the popular impression of rock stars or the hippies of the late sixties and early seventies. Unfortunately, as one scientist noted, the research to confirm this hypothesis would require some politically successful volunteers willing to "take the cure." To the best of my knowledge, so far not one politician has decided to heed the call.

There are many cases where dogs have attempted to mount humans. Since we now understand that mounting behavior is most typically a statement of dominance, it should be clear that a dog which has grabbed your knee and is merrily thrusting away is not saying, "I love you," nor is it simply trying to be "amorous." When dogs mount human beings, it is virtually always an attempt to express their feelings that they are dominant. In effect, they want to be leader of the pack. This kind of "talk" from a dog is not permissible. It must be stopped to maintain the pack hierarchy, which should always put humans above canines in dominance.

How do you stop dogs from mounting people? Since mounting is a sign of social dominance, you must assert some form of dominance or take away the social value of the act. The easiest way to exert dominance is through basic obedience training. People are often amazed at how quickly this kind of behavior can be diminished by simply taking a basic beginners' dog obedience class. Since the nature of training involves teaching the dog to respond to your commands, you are expressing your dominance, and dogs don't mount individuals they feel are dominant. On the odd recurrence of the mounting behavior, the dog is told emphatically, "No," and then quickly moved off its target, placed in a controlled "sit" or "down," and told to "stay" for a minute or two. If the commands are enforced, the human's dominance is reestablished in a gentle manner, and the mounting behavior should disappear.

Sometimes, especially with meek and mild owners and large and dominant dogs, this behavior can be quite persistent. In such cases, I have found that the best way to treat this is by attacking the "social" component of social dominance. This involves avoiding any physical contact with the dog when it displays the mounting behaviors. Physical contact and social attention can be very powerful rewards. Simply leave a short leash snapped onto and trailing from the dog's collar. Every time the dog tries to mount you, your kids, or any visitors, you should take the leash and lead him to a quiet room, where you can close the door and physically isolate him from any contact for about three minutes. After this "time out," the door is opened for him to return to human contact, without fuss or any word at all.

I once described this method that I use to stop mounting behavior to a woman who owned a fox terrier named Tracker. Although the dog was under control when her 180-pound husband was around, the moment he left the house, Tracker would try to mount her, and would persist in this activity many times during the day. The first day that this isolation procedure was started, I got a phone call from her.

"This isn't working. I had to isolate Tracker around twenty-five times today."

"Keep at it," I told her. "Tracker has had more than a year of successfully expressing his dominance and this might not be an overnight solution."

A few days later, she called again. "Tracker is doing it a lot less— maybe a half a dozen times a day. Unfortunately now I have a new problem. He's humping the pillows on the sofa!"

"That's what we call displacement behavior," I reassured her. "Since he's having less success with you, he's looking for something to dominate, even if it's only a pillow. Just remove any objects that he tries to mount, and whenever you see him mounting an object, treat it as if he were mounting you and simply isolate him for the usual three-minute period."

On each day the number of mountings and isolations was reduced, and after nearly three weeks they had been reduced to zero. Remember, mounting is a behavior designed to communicate social dominance for the dog. If the dog is immediately removed from members of his pack whenever he tries it, then it becomes a useless bit of communication because it results in having nobody around to dominate. Since canine communication is designed to provide useful and desired results for the dog, signals that produce unwanted effects will simply tend to fade away.

15

Signing and Typing

When we first started to consider the issue of dog language, we began with an idea that early researchers brought with them. In studying animal productive language, they always seemed to come back to the question of speech. Their basic argument had always been that if an animal can't speak, in the form of producing meaningful, humanlike sounds, it doesn't have real language. We have now seen that dogs do communicate, but much of their communication is in the form of signs and signals rather than carefully molded speech sounds. Perhaps if we accepted the possibility that the productive language of dogs or other animals can be in the form of body language, or gestures, we could show that they do have greater linguistic abilities than researchers have previously given them credit for—maybe even at a level of complexity that linguists would accept as "true language."

The first thing we should make clear is that this is not such a revolutionary concept. Language is possible without speech sounds. Take the case of the communications systems used by deaf people. Obviously, a deaf person may never hear others speaking, since he or she simply can't perceive the sounds associated with ordinary conversations. Yet such people can learn a language based upon gestures. In the United States, deaf people usually learn American Sign Language (normally referred to by its initials, ASL).

Is this complex system of gestures really a language? Well, although it is not simply a straight translation of any known language, it certainly has all of the components that we expect in a language, including grammar. Furthermore, ASL is not just a matter of pointing; it can be used to express ideas and can describe events that occurred in the past or may occur in the future. It can also be used to characterize and discuss objects not physically present. Complex stories can be told in ASL, just as in any spoken language.

ASL can be learned in the same casual way that spoken languages are learned by children. Babies born to deaf parents who speak ASL simply learn ASL. The baby itself may not be deaf, but it will learn the gestures from informal interactions with its parents, rather than through some form of explicit instruction, much the way that hearing children who grow up in an environment where they are surrounded by spoken language learn to speak the language their parents do. These babies will also go through the normal language development stages, and they will even babble, but not vocally—they will babble with gestures. Clearly, then, language does not have to issue from the mouth; it can also come from the hands or any other part of the body that can be used to make gestures.

Once we have freed ourselves of the idea that animals must speak and "sound like people" if they are to be credited with having any kind of language, then we can look at animal language in a more innovative way. We have seen that animals do have body language and are capable of gestures. Some animals might not be able to make gestures as complex as those found in ASL, but others can. Furthermore, once we have shed the chains of the spoken word, we can provide help for those animals who might not have very good muscle control by substituting some technological solutions for their limited ability to gesture.

When a new breed of researchers began studying animal language, dogs were not the first species they chose for study. Following the lead of early researchers, the chose the animals that were closest to humans, namely apes, in the hopes that success would be more likely. We have already seen that chimpanzees reared in typical human households and given the same language instruction as children don't learn to speak many meaningful sounds. However, as early as 1925, the primate behavioral psychologist Robert Yerkes speculated that apes might have a lot to say, but simply didn't have any way to say it. He suggested it might be possible to teach them some form of sign language. This suggestion would not be carried out until 1966, when Allen and Beatrice Gardner of the University of Nevada came onto the scene. They took advantage of the fact that apes have very flexible hands, which can form many gestures.

The Gardners obtained an approximately one-year-old female chimpanzee named Washoe. She had been captured in the wild after being raised the first few months of her life by her mother. Washoe was actually housed in the Gardners' backyard, an area of about 5,000 square feet. She lived in a completely self-contained house trailer, which provided for her toilet, kitchen, and sleeping needs. During the four years that she lived there, the researchers communicated with Washoe only in ASL. It was hoped she would learn much of the sign language from simple exposure; but there would also be some formal teaching sessions—a sort of primate school to learn ASL.

A member of the research team was with her during all her waking hours. They would chat with Washoe in sign language as they engaged in a variety of activities designed to keep her interested and active. Washoe went on frequent outings in the nearby community and had many visitors. She also played, climbed trees, and used the playground equipment in the Gardners' backyard. Interspersed were the formal sessions during which she was instructed in ASL. She was often taught by simply having her imitate gestures that the teacher made. At other times it was necessary to actually shape her hands to make the various signs. In order to encourage Washoe to use these signs, she received treats as rewards when she made correct signs for objects or situations.

Washoe did begin to learn ASL and even went through some of the early language learning stages observed in humans, such as babbling with gestures. She eventually learned 132 signs.[1]

Roger Fouts, one of the researchers working with the Gardners, later continued the study of ape language after he took Washoe to the Primate Center at Central Washington University.[2] He felt that the most compelling evidence that Washoe was using language the way human children do was what he called "spontaneous hand chatter." She would sit on her bed and sign to her favorite doll, the way that young children talk to their toys. Once, he saw her sneaking into a room which she knew was off limits, and he noticed that she was signing "Quiet" to herself.

Watching Washoe being tested for her knowledge of language produced some fascinating observations. For instance, when she made mistakes, she tended to make the same kinds of errors that children make. Her errors were based on confusion of meaning, rather than confusion of the sign for the word. Occasionally, she signed "dog" when presented with the picture of a cat, or "brush" when shown a picture of a comb, or "food" when presented with a picture of meat. She even learned how to correct her own use of words. Once, she signed "that food" when looking at a picture of a drink in a magazine. Then she looked at her hand, gave an expression of disgust, and changed the sign to "that drink." This is much the way that children often correct themselves when speaking by saying, "No! I don't mean that! I mean . . ."

In addition to learning single ASL word signs, Washoe learned to string signs together into two-word and occasionally three-word sentences. She could ask for things, "Give me apple," or, "More banana," and could describe objects: "Apple red" or "Ball big." She would request activities: "You me tickle," or describe intended destinations such as "Go out," when she was leaving the room, or, "In down bed," when she was preparing for sleep. She could determine complex relationships, such as ownership, as when she answered the question, "Whose hat is this?" with "Roger's hat," or, "Whose ball is this?" with "Washoe's ball."

Since Washoe, a number of other chimpanzees have been taught ASL. The similarity of their language use and structure to what we might see in a two-and-a-half to three-year-old child is remarkable. They will occasionally invent new labels using existing signs, such as calling watermelon "drink fruit" or a swan "waterbird." In one instance, a chimp bit into a radish and spat it out, signing, "cry hurt food." If no appropriate signs exist that can sensibly be used to label an object, they have been known to create one. There is an interesting case of this which occurred when Washoe invented a sign for "bib" that involved tracing the outline of a bib on her chest. The Gardners had wanted her to use the word "napkin," since that was what was often used as a bib, and so they insisted that Washoe use that sign. A month or so later, some ASL-speaking deaf children at the California School for the Deaf were watching a film of Washoe. When they saw Washoe signing "napkin" for "bib," they informed the researchers that that was the wrong sign. They then demonstrated the ASL sign for "bib," which involved tracing the outline of a bib in front of one's chest—almost identical to the sign that Washoe had invented for herself!

Apparently, chimpanzees can even create and use profanities. This was first seen after Washoe had been transferred to the Institute of Primate Studies in Norman, Oklahoma. There she lived in a large compound with other chimpanzees and other monkeys. Observation of her behavior showed that she continued to use sign language and even taught it to other chimpanzees in the enclosure, much the way that adults try to teach language when they interact with young children, or with individuals who don't know the language they are speaking. Up to that time, Washoe had been using the sign "dirty" to refer to feces or soiled items. After she had gotten into a fight with a particular rhesus monkey, she labeled him a "dirty monkey." Since that time, she has regularly used the sign "dirty" to describe people who refuse her requests. Just like humans, she has learned how to swear!

Roger Fouts later reared Washoe in a family setting with a number of other young chimpanzees. He and his wife, Debbi, videotaped about forty-five hours of random chimpanzee conversations. They found that, similar to a human family, they chatted as they engaged in everyday activities. They signed to each other while they played games, shared blankets, ate breakfast, and when they got ready for bed. They would even resolve difficulties by resorting to sign language. If two young chimps, Loulis and Dar, got into a fight, Loulis blamed Dar for the problem. He pointed to her and then signed, "Good good me." Washoe would then come over and discipline Dar. Eventually, Dar caught on. When he saw Washoe coming over, he would run over and sign frantically, "Come hug." Washoe would then relent and scold Loulis instead, telling her to leave the room with a "Go there," while pointing to the exit.

Chimpanzees are not the only non-human species that can learn ASL. An orangutan was taught more than fifty gestures, and the psychologist Francine Patterson has taught a lowland gorilla named Koko over 300 gestural signs. Like Washoe, Koko has learned to use profanity, but in addition she will occasionally use her sign language to tell lies, if she thinks she can get some kind of reward by doing so.

There have been some skeptics who doubt that any of this is true language. They point to the fact that most of the ape language is in the form of a request for something, and argue that the animals might simply be rote-learning a particular gesture to get a reward. Thus, they argue that if a dog responds to the command, "Sit!" by moving its body into a sitting position, and then gets a treat or a pat on the head, it has demonstrated a learned association between a sound and a body movement, but has not learned the meaning of the word "Sit." Similarly, a chimpanzee might respond to the signed question, "What do you want now?" by signing, "Give me apple," without any conceptual notion of the meaning of language or sequence, but only knowing that this particular sequence of hand movements gets a reward.

There are several points which make this argument less convincing. One is the fact that much of human language, although not expressed as a request, actually serves as a request for something, depending upon the context. Take the simple phrase, "My feet hurt." This sounds like a description of an existing condition in the world and doesn't appear to be a request, as in, "Give me apple." However, there are many situations in which "My feet hurt" is correctly interpreted by a listener as a request. In a doctor's office, this statement would serve as a request for treatment and relief of the pain. When hiking along a trial, it could be interpreted as a request for a pause to rest. When leaving work, the same phrase might be interpreted by a friend as a request for a ride home. And when entering one's home, it might simply be a request for attention, a hug and a kind word from a loved one.

There is further evidence that animal language is not simple rote learning, but has the characteristics of human language. In human language, we can express the same idea using a number of different word sequences: "The boy hit the ball," "The ball was hit by the boy," "What the boy hit was the ball," "It was a ball that was hit by the boy," and so on. Each of these sentences is different, but each communicates the same content. The same process held for Washoe. For instance, when she was confronted with a locked door, she would vary her sentences. "Give me key," "Open key," "Key in," "Open key please," "Open more," "In open help," and "Open key help hurry," are among the thirteen different phrases she was recorded using for this one situation. If we were dealing with only rote memorization of a phrase to produce a given outcome, then one phrase, once rewarded, would have been used over and over again, with no need for any variations.

Some of the analyses of video-taped conversations of the chimpanzees' everyday life also suggest more than simple learned requests. The chimps would often sit and just chatter to one another about events of the day, and what seemed to be on their minds. When they discussed their favorite food, it wasn't to get food, since there were no humans around. They just seemed to comment on it. One might say, "Apple good," and the other would dispute his choice of food with, "Banana good." Then they might go on to discuss which foods they disliked, all with no food in sight. One might notice a person walking past the window carrying a coffee cup and comment, "Coffee," while another (who found coffee too bitter) replied, "Coffee bad." The vocabulary is very simple, and the sentences are very short, but there appears to be a real attempt to use signs the same way that young deaf children use their sign language.

From a researcher's point of view, although these results seem impressive, there are some problems with using ASL as evidence of language in apes. There is the possibility that the observer, who is conversing with the chimp, may be overinterpreting and reading too much meaning into the animal's responses. The listener may even, somehow unconsciously, be leading or controlling the chimpanzee's behavior to create the appearance of more language ability than the animal actually has. For this reason, some researchers have used a different technique, which is tantamount to teaching the apes to read and write.

The first person to attempt teaching a graphic language to apes was David Premack, who began his work with the University of California and later continued his studies in Pennsylvania.[3] His first student was a six-year-old laboratory-reared chimpanzee named Sarah. Instead of written words, he used plastic pieces of differing colors and shapes, each backed with metal so that they would stick to a magnetized board. The shapes were quite arbitrary and had no relationship to any objects they represented. Furthermore, many of the words were quite abstract, such as "no," "not," or even "if . . . then." Sarah learned how to "read" these shapes as though they were words. Through simple learning procedures, she was then taught to write her responses by selecting shapes and assembling them into answers to questions or requests for various things. She has learned around 130 terms, which is around the same number that Washoe learned in ASL. Furthermore, Sarah can put these symbols together to "write" fairly complex sentences involving trading and hypothetical situations. For example, she might write: "Sarah give Mary apple if/then Mary give Sarah chocolate."

This work has been extended and placed under greater experimental control by Duane Rumbaugh and Sue Savage-Rumbaugh, working at the Yerkes Laboratories of Primate Biology just outside Atlanta, Georgia.[4] They hit upon using a variety of chimpanzee that has turned out to be a language wizard in comparison to other animals. This is a rare and endangered species, *Pan paniscus,* sometimes known as the pygmy chimpanzee—a misleading label because these animals are almost as large as the more common chimpanzee. The species is also sometimes known as the bonobo. The language learning system that they use is somewhat similar to Premack's, only it is completely computerized. A keyboard containing seventy-five to ninety keys is used. Each key is marked with an arbitrary symbol. When the key is pressed, it lights up and its symbol appears in sequence on a screen, so the chimpanzees can keep track of their symbol sequences as they "write sentences."

The performance of these bonobo chimpanzees has really been quite impressive. They occasionally use the symbols to name and describe objects that they are not asking for but simply are viewing at that moment. They also sometimes use their typed symbols to describe events that occurred in the past, as when one bonobo explained a cut on his hand by describing how his mother had bitten him. Sometimes they will make very creative requests, such as asking others to do something to someone else; one bonobo asked one researcher to chase another so that he could watch them.

These chimpanzees were not surrounded by ASL, but rather, lived in an environment where English was spoken by the researchers. The researchers would speak to the bonobos when they were teaching them new symbols, so that a sound might also be associated with a sign, and they also would speak to the chimpanzees in a casual fashion about whatever was being done at the moment. Just like young children growing up surrounded by a particular spoken language, these chimpanzees developed receptive language abilities, and have achieved a very good comprehension of spoken English. Their language ability is sufficient for them to be able to respond to commands that use combinations of words they have never before heard used together. For example, one might be asked to "Take the key and put it in the refrigerator." Although they might understand each of the individual words, this sentence represents a new concept they have never before been presented with. Nonetheless, the animals still respond accurately.

The similarity between the way in which the bonobos seem to have learned language and the way in which human infants learn language is remarkable. Much of this occurs simply by observing others using the language and through normal social interactions involving the language. One bonobo, Kanzi, learned how to type out sentences by watching his mother being taught when he was very young. The researchers actually gave up on his mother, since she appeared to be very slow to learn and not particularly bright. However, once his mother had left the lab, Kanzi showed that he had developed good language skills, not only in his receptive language but also in his productive language. He showed that he had already learned how to correctly use the keyboard to request items of food, and then demonstrated that he had also learned how to request activities, such as watching TV, playing games, visits to friends, and so on. Perhaps most surprising is the fact that Kanzi also uses the keyboard to announce his intentions, as in: "Kanzi eat apple then . . . next go bed." Some researchers have suggested that this level of linguistic performance is almost as good as that of a human three-year-old child.

Although these studies of ape language abilities are promising, their application to the language abilities of dogs may be somewhat limited. The studies do tell us that some forms of body language (which is what signing is) can be learned by animals more easily than vocal language. The obvious reason that the ASL studies are not such a hopeful route for dogs is that not only do dogs have limited vocal control to shape the sounds of words, they also have limited manipulative ability. They don't have the dexterity of apes and cannot shape gestures. Even if they had the required agility in their paws, they have no fingers to form the shapes needed for ASL or any of the complex sign languages. Dogs can learn to poke at things, and perhaps draw them closer or hold them down using their paws, but the dog's basic means of manipulating the world involves its mouth and jaws.

However, the more recent computerized studies involving the use of keyboards do have some promise. A dog could be trained to push a key with its nose, or perhaps even to place a paw on a specific key. If, somehow, these responses were associated with symbols, perhaps some aspects of human language could be taught to the dog.

This brings us to the story of Elisabeth Mann Borgese and Arli. These events predate the symbol manipulation experiments of Premack and the Rumbaughs, and even came before the Gardners began to teach Washoe sign language. Elisabeth was the youngest daughter of Thomas Mann, the German novelist who won the Nobel Prize for Literature in 1929. She was a writer, environmentalist, and also an ardent student of animal behavior. In October 1962, Elisabeth began what would be a three-year experiment, in which she would attempt to teach her dog Arli to read and write. She would not be using some arbitrary new communication system, but rather, human language. She had selected Arli as the brightest of her four English setters, and the most likely to benefit from schooling. By the time she was finished, Arli would be taking dictation and typing the words that she spoke to him.

Her teaching methods began quite simply. She used plastic cups, which could be covered by plastic saucers. Each saucer had a symbol on it. The dog's job was to decide which symbol was correct, and then to knock that saucer off the cup. If he was correct, he would find a little bit of food in the cup, and this served as a reward. Thus, she began with saucers marked with either one or two large black dots. If she said the word "one," the dog had to select the saucer with one dot, while "one-two" was the verbal signal to select the one with two dots. This first step in language learning took four full weeks of training.

In order to get the dog to pay more attention to patterns, she went on to teach him to discriminate between different drawn symbols, such as a plus sign versus a circle, or a triangle versus a square. Once a number of such discriminations between pairs of patterns was learned, she moved to multiple-choice learning. Now the dog was presented with three cups marked with one, two, or three dots. The added three-dot pattern had to be selected if she gave the new verbal command, "one-two-three." The order in which the patterns were presented was always mixed up, and changing, so the animal really did have to look at the patterns and count the dots to find the cup with the correct number. Arli was not a natural whiz at mathematics, but after three months of daily training, the dog had learned to count to three.

Arli's rate of learning actually sped up after a brief vacation from schoolwork, and it took him only one more month to learn to count to four, and also to tell the difference between two words: "DOG" and

"CAT." Elizabeth explained, "This is the way in which animals 'read': you say 'dog' and he knocks down the saucer marked DOG; you say 'cat' and he knocks down the saucer marked CAT."

After another few weeks, Arli could count all the way up to six and read the words DOG, CAT, ARLIE, BIRD, BALL, and BONE. He had also learned to pick out the larger of two numbers that were presented to him, although Elizabeth admitted that this took "many days and weeks and involved thousands of errors, disappointments and setbacks."

The next task set for Arli was to learn to spell. Elisabeth now took a familiar word image, like DOG, and presented it as three saucers marked D, O, and G, respectively. Arli had to knock the saucers down in the proper order to spell out the test word, even if the saucers themselves were ordered in some different sequence like ODG or GDO. Regardless, he had to pick out the D first, then the O, and finally the G. Eventually, Elizabeth introduced mixed sets of letters, like DCOAGT, and Arli had to pick out the words DOG or CAT on command.

Training didn't always go smoothly. When Arli was tired or found a task particularly difficult, he would sometimes just stand there, with a particularly befuddled expression, waiting to be helped. At other times he would just knock down saucers at random, like a college student taking a multiple-choice examination he hasn't studied for. He figures that answers marked at random might, by chance, be correct, while no answer at all will certainly result in the lowest possible score.

After Arli was fairly reliable at picking out words on saucers marked with letters, he was transferred to an electric typewriter. This was a keyboard with twenty-one letters and a space key, which could be activated by Arli. All he had to do was push his nose against the appropriate key. There was no special computer monitor for Arli to gaze at, since computers were not yet common, so the only way that Arli could monitor the sequence of letters he typed would be by looking at the letters as they were printed on the page with each keystroke. To help him, Elisabeth thoughtfully placed a magnifying glass in front of the carriage, which should have allowed Arli to see enlargements of the letters and words that he was typing. This, however, turned out to be dispensable. There seemed to be no way of drawing Arli's attention to the written sheet, nor of connecting the finished typescript with the actual activity of typing. As far as Arli was concerned, he was through when he had typed a word, or perhaps better put, "acted out" a sequence of letters. The finished manuscript or typed page was then only something that might be good for chewing.

Arli learned very quickly to write words, including: ARLI, PLUTO (another of Elizabeth's dogs), DOG, CAT, BIRD, CAR, ROME MEAT, BONE, EGG, BALL, GOOD, BAD, POOR, GO, COME, EAT, GET, AND, NO. These were dictated phonetically to him, with careful and prolonged enunciation, as in AAAAA-RRRR-LLLL-EEEE. Unfortunately, Arli seemed to attach virtually no meanings to the words. It seemed more as though he was "taking dictation" rather than learning to read. After a while he had learned seventeen letters, around sixty words, and could type whole phrases, such as *good arli go car and see a bad dog,* without errors. Elisabeth felt very proud of his performance for that year. In fact, she was so confident of his abilities that she let him type out her Christmas cards.

Did Arli really understand the meaning of anything that he was typing? Elisabeth was not completely sure; but there was one incident which gave her some hope. Elisabeth had been traveling with Arli and he had developed some gastric problems. These made him listless at work. One day she called him to the typewriter. He was fairly lethargic and really didn't seem very interested as she dictated "g-o-o-d d-o-g g-e-t b-o-n-e." He seemed as if he couldn't have cared any less. As she stood there, he finally moved to the typewriter and put his nose to the key marked "a." Although she had dictated no "a," she thought she would let him do it anyway. Arli proceeded to type, without any prompting and with all the spaces correct: *a bad a bad doog.* Elisabeth had great hopes that the breakthrough had finally come and she was on the verge of true written communication with her dog.

As soon as Arli had recovered fully, she decided to try a new experiment. Elisabeth would let him type without any dictation. She would let him type whatever passed through his mind (or nose). After looking at the result, she decided that he was actually writing poetry, not prose. Arli wrote in continuous lines, but Elisabeth, while carefully maintaining all the spaces between units, divided the long lines into short ones, to emphasize the rhythmic quality of the text. She also completed each "poem" by adding a title.

It turns out that Arli's use of recognizable words in his poetry is a bit sparse. My favorite poem of his is

> BED A CCAT
> cad a baf
> bdd af dff
> art ad
> abd ad arrli
> bed a ccat

Elisabeth submitted some samples of Arli's work to a well-known critic of modern poetry, without telling him the poems were written by a dog. The critic responded that "the poems are charming. I think he has a definite affinity with the 'concretist' groups in Brazil, Scotland, and Germany. Has he been in touch with them?" He then went on to suggest that, if his work was supported and nurtured, Arli might eventually evolve to the stature of the American poet e e cummings, "who is also writing poetry of this type at present."

Elisabeth could have done that nurturing, but decided against it. As she later wrote:

> By rewarding only correct words while leaving to Arli the task of choosing freely from among his by now rather large vocabulary of words and word combinations, I could have trained a preference for real words as against causal sequences of letters. After a while the poems should have looked still more human and less concretist.
>
> But I abstained. Spontaneous typing is hard on Arli's nerves. He gets restless. He begins to hit the keys with his paw. He begins to whine and whimper and yelp. "How should *I* know what to do?" he seems to say, "Dictate! For heaven's sake *dictate!*"[5]

Sadly, the written word appears not to be the best medium for the expression of ideas by a dog. Arli seems much like a secretary of mine who was once transcribing some old reports and putting them into my word processor. She had been working on this task for several days when I wandered by her desk and asked her whether she found the material interesting.

"I can't say," she replied. "I just type it. It's not like I read it or understand it."

So it seemed to be with Arli, the secretarial English setter.

16

Scent Talk

We can learn to understand and even to communicate with dogs through a knowledge of their language, but this can only occur when their signals are designed to be received by senses that work well in human beings. We understand the messages that the dog sends by sound, and can read the displays on his face, read his touch signals, and interpret his communication "dances" and body postures by sight. There is, however, one important channel of canine language that will remain forever a mystery to most humans, and that is the language of scent.

The average person has around 5 million scent receptors in his or her nose, which puts us in the lower third of mammals in our smell sensitivity. The average dog has around 220 million receptors in its nose, which potentially makes its sense of smell forty-four times more sensitive than ours. Furthermore, evolution has designed the dog's nose to make the maximum use of that multitude of scent receptors. To begin with, the canine has mobile nostrils that help it determine the direction of the scent. Even his sniffing patterns are different from those of humans. The dog does not need to fill up his lungs as he continuously brings the odor into his nose in bursts of three to seven sniffs. The dog's nose contains a bony structure inside that humans don't have. Air that is sniffed passes over this bony shelf, and many odor molecules stick to it. The area above this shelf is not "washed out" when the dog exhales, and this allows those scent molecules to remain there and to accumulate. When a dog breathes in normally, the air passes through the nose and continues down into the lungs. Sniffing, however, leaves the air resting in the nasal chambers so that the scent can build in intensity. This means that even incredibly faint odors can be detected.

Just how sensitive the dog's nose is has been demonstrated by the U.S. Army, which has used dogs to detect land mines. Concern about clearing mines has been growing over the years, since many mines are now made with plastic parts (except for the fuse contacts), which makes it difficult to locate them with metal detectors. A report from the Army Research and Development Center in 1985 concluded that there were no mechanical or electronic devices as effective as a dog in detecting mines, booby traps, and explosives. Furthermore, the dog's ability even seems rare among animals, since they also tested badgers, coyotes, deer, ferrets, red foxes, various types of pig (including a variety of the wild boar called a javelina), raccoons (and their South American cousin the coati), skunks, opossums, and beagle/coyote crosses. None did as well as the dog.

The army researchers gave the dogs some incredibly difficult tasks to perform during these tests. Researchers buried mines and left them there for weeks or months before the dog was required to find them. They poured oil on the ground and set it on fire to cover the smell; they sprinkled the ground with live and spent ammunition to provide distracters. Nothing seemed capable of defeating the dog's nose.

Dogs start off their lives operating almost exclusively by smell and touch. It is the heat of their mother that first attracts them, but the blind newborn puppies must use their sense of smell to find their mother's teats in order to suckle. Within a few days, they can distinguish their mother's smell from all others. Simply bringing their mother into a room when they have been isolated from her will tend to quiet the puppies, even if she is brought in silently and they cannot see her. Her scent is the smell of safety and comfort.

Dogs' sensitive sense of smell is always producing new surprises. A while ago I spoke with Richard Simmons, who was a research associate working on a project supported in part by the National Institutes of Health in the United States. He told me a story about Marilyn Zuckerman of New York and her Shetland sheepdog, Tricia. Tricia had developed the annoying habit of sniffing or nuzzling Marilyn's lower back whenever she sat down and the dog could reach it. Marilyn's husband noticed there was a dark mole on her back that Tricia seemed to be interested in. It seemed odd that the dog cared about this mole, but since it caused no discomfort, Marilyn simply ignored it. Then, one spring day, Marilyn was lying facedown on her balcony in a bathing suit, enjoying the sunshine. Suddenly, she felt teeth on her back. Tricia had apparently decided that that mole shouldn't be there and was trying to remove it. Tricia bit at it so hard that Marilyn gave a yelp of pain and jumped up.

It was at this point that Marilyn's husband suggested there must be something really odd about that mole if it was bothering the dog so much. Marilyn was going to the doctor for another reason, and more to satisfy her husband's curiosity than anything else, she pointed out the mole. Before the day was out, Marilyn was at the Cornell Medical Center, where the mole was diagnosed as skin cancer—actually a virulent and dangerous form of melanoma, which can be fatal if not caught early enough. Tricia's early warning probably saved Marilyn's life.

Simmons told me, "It was because of stories like that that we began testing dogs' diagnostic abilities. Our preliminary data suggest that dogs can detect melanomas and several other types of cancer well before there is any other indication of a problem. We believe these cancers must give off some form of odor which the dog's nose can pick up. Some dogs will show agitation the moment a person with cancer enters the room. It may well be that, someday in the future, inspection by a dog may become a routine part of cancer screening."

Now, although all dogs seem to have an acute sense of smell, we can't say that all dogs are created equal in their scenting ability. Male dogs seem to do better than female dogs, perhaps because males are more competitive and more sensitive to scent marking by other males nearby. There are also some breed differences. Dogs with pushed-in faces, such as pugs or Pekingese, don't do as well as a group, probably because this face shape often leads to respiratory problems which may affect the normal flow of air through the nose. The best noses are found in some of the hounds, with the bloodhound probably the all-time champion. It has been scientifically shown that a bloodhound doesn't lose the track it is following even if its quarry puts on rubber boots or jumps onto a bicycle as part of an escape strategy.

Just how sensitive the various breeds are to scents was tested in part by John P. Scott and John L. Fuller at their lab in Bar Harbor, Maine.[1] They put a single mouse in a one-acre field and released some beagles. These smell-sensitive dogs took just one minute to find the tiny rodent. When the same test was used with fox terriers, it took them close to fifteen minutes to find it. A group of Scottish terriers were actually never successful at locating the mouse by smell. One actually stepped on it, and was finally attracted to its presence by the squeaking sound it made. I suppose that's why we never use Scotties to track missing fugitives and lost children.

I think it is fair to say that dogs perceive the world differently from people. For dogs, reading scents is the equivalent of reading a newspaper. The special scents that dogs and other animals produce for the purpose of communication are called *pheromones,* from the Greek words *pherein,* meaning "to carry," and *horman,* meaning "to excite." The original belief was that these smells simply told male animals when females were in season, and then served as a means of exciting them to track the females and mate with them. Today, we know that these personal chemicals carry a lot more information than sexual readiness. Different hormones are secreted when an animal is angry, fearful, or confident. Some chemical signatures identify the sex of the individual, and others tell us how old the dog is. There is a lot of sexual information as well, such as where the female is in the estrus cycle, if she is pregnant or having a false pregnancy, and even if she has given birth recently.

If reading scents is, for dogs, the equivalent of reading a written message, then the canine equivalent of ink is urine. Many of the pheromone chemicals are found dissolved in dog's urine. That means that a dog's urine contains a great deal of information about that dog. Sniffing at a fire hydrant or a tree along a route popular with other dogs is a means of keeping abreast of current events. That tree is really a large canine tabloid, containing the latest news items in the dog world. While it may not contain installments of classic canine literature, it certainly will have a gossip column and the personal section of the classified ads. When my dogs are busily sniffing at a favorite spot or tree on a city street frequented by other dogs, I sometimes fantasize that I can hear them reading the news out loud. Perhaps this morning's edition goes: "Gigi, a young female miniature poodle, has just arrived in this neighborhood and is looking for companionship—neutered males need not apply," or, "Rosco, a strong middle-aged German shepherd, is announcing that he is top dog now, and is marking this whole city as his territory. He says that anybody who wishes to challenge this claim had better make sure their medical insurance is current and paid up."

The biggest difference between canine and human reading is that humans are allowed to finish the entire piece. Many dogs only get to "read the headlines" before they are hauled away by the pull of their leash. This occurs because many owners feel that the process of sniffing where other dogs have left their urine marks is unclean and disgusting. Some unenlightened dog owners may even discipline their dogs for trying to keep up on the neighborhood news.

The reason that fire hydrants and trees are popular places for urinating is that male dogs prefer to "mark" vertical surfaces. Having the scent above the ground allows the air to carry it much farther. Perhaps the most important reason for using elevated and vertical surfaces as the target is that the height of the marking tells the neighborhood something about the size of the dog making the mark. Remember that among canines, size is an important factor in determining dominance. Since dominance seems to be more important to males, they have developed the habit of leg lifting when they urinate so that they can aim their urine higher. Also, the higher the marking, the more difficult it is for other dogs to mark over it and obscure the message.

Some dogs try so hard to make their urine marks as high as possible that they almost fall over in the attempt. I have actually seen a rather bizarre example of an attempt to produce extremely high urine marks. This involved a Basenji, the small African sight hound that is considered to still be very close to the African wild dogs in many of its behaviors. This particular Basenji, a strong unneutered male named Zeb, had adopted a pattern of urination that is sometimes used by wild dogs. He would aim himself toward a tree and then run directly at it. Zeb would leap when he was near its base so that his hind feet were essentially walking up the tree. His momentum typically carried him five or six feet up the trunk. Finally, he would flip over at the top of his run, so that he landed on his feet having performed a perfect loop. The real purpose of this trick was shown by the fact that he performed his acrobatic somersault with a continuous stream of urine flowing. Of course, this left a stripe scent that rose well above those put out by any other dogs in the vicinity. I often wondered what his canine neighbors thought as they read Zeb's announcement. "Hmmm, I think we may have a King Kong–sized dog living nearby."

Although it is usually males who lift their legs, it is not unusual for a female to do so also. This seems to depend somewhat on her self-esteem and confidence. More dominant females are much more likely to lift their legs when urinating, while those who are more unsure of themselves are less likely to do so. Sexual status also plays a role. Spayed females are much less likely to leg-lift, although dominant females still sometimes do so even if they are no longer fertile. Environment also plays a role. If there are a lot of sexually active females around, any given female is more likely to lift her leg when urinating. In Denmark, where few female dogs living in the city are ever spayed, you are more likely to see females lifting their legs than you would in the United States or Canada, where it is usual to have most of the city dogs spayed.

Dogs and wolves often use urine to mark their territories. Roger Peters, a psychologist and wolf researcher, studied these markings.[2] He found that the wolves use urine to mark the perimeter of their territory so that, in effect, they live inside a region ringed in urine. They also use urine to mark certain pathways which are important to them. This means that, for the wolves, these urine scent posts form a map of their region, telling visitors about the inhabitants, and reassuring members of the pack that they have returned to familiar terrain. Wolves and dogs will mark more frequently as they move out of their own territory; it has been suggested that this has the same purpose as a human's blazing of trees, so that he can find his way back home after the journey.

Wolves and dogs use not only urine but also their stool to mark their territory and important regions within it. The anal glands of canines give a particular signature to fecal deposits. These identify the individual who is leaving the droppings as well as marking the place where the dung is left. Dogs are very particular about these landmarks. This explains what appears to humans to be a meaningless and complex ritual that dogs engage in before defecating. Most dogs start with a bout of careful sniffing of a location, perhaps to set an exact line or boundary between their territory and someone else's. This behavior may also involve finding some small degree of elevation, such as a rock, a fallen branch, or the lower leaves of a bush, to leave the feces on. Again, elevation will allow for dispersal of the scent over a maximum range.

Deposits of stool and urine are so important in marking the region that dogs and wolves often leave visual and olfactory signposts to make sure that other visitors to the area will find them. Most male dogs, and a reasonable number of female dogs, will scrape or scratch the earth with their hind legs after leaving their scent marker. Because the earth flies backwards from the scraping, and some may actually fall on the place where the urine or feces were left, some people concluded that this activity was designed to cover the smell and to hide the feces. Cats scrape with exactly that intention; dogs do not. Another hypothesis was that perhaps dogs were trying to spread the fecal matter around. If that were the case, then generations of dogs have developed very bad aim, since their scraping behavior seldom actually disturbs the deposit.

Recently, we have come to understand that scratching the ground is actually a visual sign, which points to a scent marking. If other dogs are in the area, they can see by the enthusiastic ground scraping that another dog has left its mark. They will usually come over and give it a sniff. They can then read the latest news and observe the proper territorial protocols.

One important factor in all of these markings is the freshness of the scent. Time and weather will erode the scent marks, and these often have to be renewed. Thus, the freshness of the scent will give visitors some idea as to the current status of this piece of geography, and whether the local inhabitant uses this area frequently. Regions which are under dispute, or which are used by different animals at different times, may result in marking battles, where every scent from "the other team" in the contested area is marked over by each competitor when they encounter it. This same behavior is seen in some regions of New York and Los Angeles, where human gangs competing for a neighborhood or "turf" announce their claims with spray-painted graffiti, only to have their "marks" covered over by another gang the next day as a sign of challenge.

Obviously, we as humans are relatively oblivious to the actual content of the messages left by dogs in their urine. There have been some cases, though, where humans have attempted to communicate something to dogs by using urine. The Canadian naturalist and author Farley Mowat wanted to assure the safety and privacy of his campsite while observing wolves.[3] He carefully urinated on rocks that marked a perimeter around his living area. When the wolves discovered his scent marks, they went around to the opposite side of each rock and marked it with their scent. Thus, each rock had a side announcing Mowat's territory and another marking the border of the wolves' territory. Mowat reports that although the wolves would often patrol along that smelly boundary, they clearly did get the message, and they respected his space.

I did hear of one researcher who was studying wolves at Isle Royale in Michigan, who claimed that he tried to repeat the Mowat experiment, but the wolves simply ignored his boundary markings. Since I don't know the full details of this attempt, I have often wondered whether this researcher's urine had nothing interesting to say, or perhaps his pattern of marking made the communication unclear. It is much like the fact that some people can tell a joke that brings gales of laughter, while others, repeating the material word for word, cannot provoke even a small smile.

I do know of an instance where human scent markings were used to communicate successfully with domestic dogs. A friend and colleague of mine at the university had a problem. His wife had decided to put in some new flower beds on either side of the front door to his house. She dug up the earth and surrounded the beds with decorative rocks. Unfortunately, the newly turned soil and the smell of new plants attracted some of the local dogs, who were digging out her flowers almost as quickly as she planted them. My friend had read Mowat's wonderful book Never Cry Wolf and thought that perhaps he could mark the rocks surrounding the flower beds with urine—sort of defining a tiny botanical territory—and that might keep the dogs away. One night, he sneaked out of the house and carefully urinated around one of the flower beds. He only marked around one, because, as a scientist, he wanted to test the effects experimentally, and this allowed him to compare the results across the two beds. Sure enough, over the next forty-eight hours, the unmarked flower bed was disturbed and partially dug up, while the urine-marked beds were not. Flushed with success (and a full pot of tea), he marked both beds this time. Since he knew that the effects would wear off in time, he renewed the marks every couple of days, and the neighborhood dogs seemed to respond as he had hoped. They would sometimes come up and urinate on the rocks that made up the boundary, but they didn't cross over and they didn't dig.

Success doesn't always come easily. A few weeks into his territory-marking program, my friend showed up in my office looking for another solution to his problem.

"It's working with the dogs, but it's resulted in other problems. I do it at night to be discreet, but this morning as I was coming to work, my neighbor stopped me. 'I know what it's like having a houseful of daughters who always seem to be in the bathroom when you need it. Since you seem to be encountering that problem a lot, you can just tap on my door instead of . . . Well, you know.'

"Even worse, when my wife found out what I was doing, she was disgusted by the whole thing. 'You don't expect me to work in those flower beds after you've just used the borders for a lavatory, do you?' So tell me, what am I supposed to do?"

I told him to take a little bit of household cleaner—one with some form of noticeable scent—and mix it with ammonia, which is one of the odorants in urine. The household cleaner was there just to give some complexity to the scent (and to convince his wife that the rock border had now been cleaned), while the ammonia was to make it all smell like some form of strange urine. I told him to put it in a spray bottle and apply it to the rocks surrounding the flower beds. He did, and the flowers have still not been disturbed, although I often wonder what kind of dog the neighborhood canines think that scent represents.

Dogs seem to consciously manipulate or play with odors. Over the years, many people have asked me why their otherwise apparently sane dog will roll around in garbage or dung or something equally offensive in its smell to humans. Several theories have been put forward to explain this behavior. One of the silliest is that this is a means of fighting parasites. The notion is that insects, such as lice and fleas, wouldn't hang around on something that smelled that bad. Unfortunately, most insects don't seem the least bit put off by bad odors on a dog.

A second theory claims that this is a means of "writing a message" to the other members of the pack. A dog or wolf will roll around in something that smells bad but may still be edible. It then returns to the pack. The other members of the group immediately pick up this scent and know that there's something that can pass for food around.

A third theory suggests that the dog is not trying to pick up odors from the stinking mess, but is actually trying to cover that smell with its own scent. It is certainly true that dogs and wolves will often roll around on something, like a stick, a new dog bed, or such, as if they were trying to deposit their scent on it. Some psychologists have suggested that dogs often rub against people to leave a trace of their scent and to mark the individual as a member of the pack, much the way that cats rub up against people to mark them with their odor.

The explanation with the best evolutionary thrust is that this smelly behavior might be an attempt at disguising the dog. The suggestion is that we are looking at a leftover behavior from when our domestic dogs were still wild and had to hunt for a living. If an antelope smelled the scent of a wild dog, or a jackal or wolf, nearby, it would be likely to bolt and run for safety. For this reason, wild canines learned to roll in antelope dung. Antelopes are quite used to the smell of their own droppings and therefore are not frightened or suspicious of a hairy thing coated with that smell. This allows the wild hunting canine to get much closer to its prey.

I have another theory, which is of no scientific merit whatsoever. Dogs, like people, enjoy sensory stimulation and may well be prone to seeking such stimulation to an excessive degree. I believe the real reason they roll in obnoxious-smelling organic manner is simply an expression of the same misbegotten sense of aesthetics that causes human beings to wear over-loud and colorful Hawaiian shirts.

Although I may have been giving the impression that dogs can extract massive amounts of social information from smells and people can extract virtually nothing, this is far from true. Human beings do produce pheromones, just as other animals do, and it would be unusual for evolution to retain such an ability if it were not used in some way. It now seems likely that we use information from the pheromones of other humans, but are often not conscious of the fact that we are picking up scent signals. Scientists have recently demonstrated that smells may also play an important role in the social behavior of people.

Many of the studies on human ability to recognize scents have involved the "smelly T-shirt" technique. Volunteers must give up use of soap, perfume, aftershave lotion, and so on for several days, and must agree to wash themselves only in clear water for the duration so that their scent is not contaminated. They are given a sterilized T-shirt to wear for a number of hours. This T-shirt is then removed and placed in an airtight container, where the scent can be concentrated, and afterward delivered to other people to sniff in measured and controlled doses. The results are quite interesting.

First of all, humans can detect their own body odor from among a set of similar scents contributed by other volunteers. People also can identify the sex of an anonymous odor donor. When we press them to describe what a man or a woman smells like, the most common descriptions are that men have a "musky" smell while women smell "sweet." In addition, male smells are often described as being strong and maybe a bit unpleasant, while the female smells are pleasant and less intense.

Human females are much better at this task than males. Not only can they tell the sex of an individual by scent alone, but also whether it is an infant, child, adolescent, or adult. Males do not have this age-discrimination ability, although some seem to be able to tell the smell of an infant from the others. Even very young babies can identify the scent of their mother's breast, and when only a bit older they can recognize their mother's body and breath odors. Parents recognize their own children's smells from the smells of others, and brothers and sisters recognize each other's scents as well.

Generally speaking, many studies seem to show that people respond to some scents almost at an unconscious level. Perhaps the behavior where humans most frequently process scent information unconsciously involves sex. There are particularly large clusters of pheromone-producing glands around the genital area. When sexually excited, both men and women often emit strong odors from this region and elsewhere. There is now a body of evidence which says that such smells may be a vital component in human sexual behavior and in interpersonal attraction. When people lose their sense of smell (a condition called *anosmia*), nearly half of them will notice a massive decrease in sexual interest and nearly a quarter of them will report difficulties in sexual performance and a great reduction in the pleasure associated with sex. In other words, these sex-related smells that have gone mostly unnoticed by us may be a necessary part of human sexual behaviors.

If that is the case, then it should not be surprising to find that perfume manufacturers are looking for pheromones to include in their products to make them "sexier." This is not a new search. For centuries, extracts from the sexual scent-producing glands of various animals have been used in perfumes. Musk is the sexual scent which comes from certain Asian deer, while civet comes from the genital area of a wild cat, and castoreum is the sexually exciting smell from beavers. These scents have been included in perfumes because they are believed to excite people. Not only do they excite the target person, they also excite the wearer, causing them to release a little of their own sexual pheromones, hence increasing the effect.

There are two important features to note here. Not only are we more sensitive to scents than we are usually led to believe, we are also sensitive and respond to pheromones when they are produced by other mammals. For this reason, perfume manufacturers can now use alpha androstenol, which is a sex-attractant pheromone from pigs that is also present in human underarm sweat.

Do such scents actually play a role in human sexual attraction? The scientific results are interesting. In this kind of research, men are given a whiff of alpha androstenol, which usually goes completely unnoticed at the conscious level. While this scent is in the air, they are shown a photograph of a woman. When asked questions about the person in the photograph, they will rate that woman as more sexually attractive than a similar woman whose photo was not seen in the presence of this pheromone. Similarly, women exposed to this same pheromone overnight seem to be more willing to initiate social interactions with men (but not with women). There is even one study which noted that a bit of this pheromone smeared on a job application affected the rating of candidates for employment. One has to be careful, though, since the effects were different depending upon whether the applicant was being rated by a male or a female evaluator. All of these effects occur, even though the scent stimulus is not consciously noticed.

If people respond to animal pheromones, even at an unconscious level, then it should not be surprising that dogs respond to human pheromones. Dogs often sniff the genital and anal areas of other dogs. This gives them some of the same odorants they would get from sniffing the urine or stool and also some additional, more sexual smells. That is also why dogs will sometimes engage in embarrassing episodes of sniffing humans when company comes. Many dogs show a desire to sniff the crotch of a person who has recently had sexual intercourse. Dogs also seem to be attracted to women at around the time of their ovulation and to women who have recently given birth (especially if they are still nursing a child). Certain medications seem to change the human scent, and perhaps certain foods do as well. When your Labrador retriever starts nudging Aunt Matilda's skirt, he is only seeking further information about her because of some interesting pheromones that she is producing. He has no idea that people consider this impolite.

Humans often overreact when a dog starts examining their body for scent messages. Take the case of Barbara Monsky, a local political activist living in Waterbury, Connecticut, who actually brought a suit against Judge Howard Moraghan and his golden retriever, Kodak, for sexual harassment. The basis for her legal action was that Moraghan often brought his dog to Danbury Superior Court. Monsky contended the dog had "nuzzled, snooped or sniffed" beneath her skirt at least three times. According to her, the judge was complicit in this harassment because he had done nothing about it.

The case was finally heard by U.S. District Judge Gerard Goettel. He dismissed the case, and in a later interview explained that "impoliteness on the part of a dog does not constitute sexual harassment on the part of the owner."

The angry and aggrieved Monsky responded by calling the judge's decision "as insulting as having a dog sniff under a skirt."

To a dog, however, this kind of behavior is no more insulting than pressing the playback button on an answering machine to see if any interesting messages have come in. The fact that one of the important human scent message centers happens to be between our legs is just a minor inconvenience as far as the dog is concerned.

While we cannot extract the vast amount of information from scents that dogs do, we do respond quite reliably to one scent message that a dog sends: humans can reliably tell the smell of a puppy that is less than nine weeks of age from all other doggy smells. This is quite automatic, and even human children have this ability. When I first got my Nova Scotia Duck Tolling Retriever, Dancer, the neighbors' kids came to see him. The three of them were between ten and twelve years of age. One of the little girls lifted him up, gave him a snuggle, and announced, "He even smells like a puppy!"

Although humans can't normally know what a dog is smelling or what it is learning from the scents around us, there are some ways in which dogs can describe what they are scenting. Some of the hounds are particularly good at this. I first encountered hounds that could tell you what they were smelling when I was in the army and taking some training at Fort Knox, Kentucky. In the countryside surrounding Fort Knox, there were many people with a real love for dogs, and I got to know a few of them. The most popular dogs in the area seemed to be hounds. The star of the region was a redbone hound named Hamilton, who had become famous locally for his ability to find and tree wildcats. I had also just learned something about bluetick hounds which sounded interesting. Both the redbones and the blueticks supposedly had been bred to "have the music," which meant that they would give a different hunting sound for different game. I didn't really believe this, but I wanted to find out for myself.

The word in the region was that there was a Baptist minister who had "the best bluetick hounds in the world." He was usually referred to as "Reverend John," or simply "Brother John." Late one Saturday afternoon, I managed to find my way to John's home.

As I approached the house, I saw two of the dogs. They were tall hounds, with dark muzzles and black-trimmed ears. Their fur was mostly white, with a fine array of black tickmarks scattered down their backs. If the sun caught those black marks just right, you could see a flash of purple-blue, which is apparently how the dogs got their name. One of the dogs, an old male, gave a preemptive bark, then sauntered down to greet me; the other, a lovely young bitch, was more reserved and stayed on the porch. Brother John had come out on hearing the bark, and waved at me.

"So you the one that want to see my dogs?"

"Yes. I hear you have the best blueticks in the country. I hear they talk to you and tell you what they're hunting."

We sat on the edge of the porch, and Brother John pulled out a couple of enameled metal mugs and a clear bottle with some amber fluid. He filled each with a healthy portion, raised his cup, and toasted, "To God's love." We sat sipping some local bourbon as he talked about his dogs.

"I been breeding blueticks for nearly thirty years. I breed them so that they got a good nose and some smarts and the desire to hunt. But I also breed them so's they can tell you what they're smelling. With me, if a dog don't sing right, it don't get bred." He pointed at the old male and continued, "Now Zeke here, he's typical of my breeding. When he's onto a rabbit, he's got a kind of a 'yip-yodel' sound. When he's onto a squirrel, it's mostly 'yip,' and when he's got a noseful of raccoon, it's mostly 'yodel.' When he's got the track of a bear, he gives a sort of a growl-bark but not very loud. If he gets a whiff of a big cat, he gives mostly a high—almost squeaky—bark. Now Becky over there," he motioned to a bitch who had now settled comfortably in the sun nearby, "she don't do bears. When she smells one, she just stands and growls and won't track. She yips and yodels like she should for the other critters, like she should. Her cat bark, though, is different than Zeke's. There's only a little tiny turn-up at the end of each bark, not the real squeak that most of my males have. The real music comes when they got a noseful of deer. I mean, they sound like proper hounds then—like bloodhounds after some lost soul or some escaped criminal. Ain't no sneaking up on a deer when they got the scent.

"Different hounds and different lines of breeding can have different hunting words. Redbones are different from blueticks, but they seem to know each other's words. I was out near Brownsville once, and Stephen was chasing a big cat with Hamilton, you know that big redbone of his. Hamilton's sound when he's hunting cat is mostly like my dogs' sound when they're hunting deer, only much more excited and broken up. Zeke hears Hamilton and takes off in the direction of his sound, but he's giving that squeaky bark of his which says cat. Maybe dogs got dialects, or maybe they just translate other dogs in their heads." He looked at me, smiled, and continued, "It could be they just make it up as they go just to keep us humans confused."

Brother John had not started the trend of breeding his dogs for their voice. For centuries, the scent hounds have been systematically bred, not just for their scenting ability and desire to track but also for the sounds they make when hunting. The hounds' baying when tracking acts as a beacon, which allows the hunters to know exactly where the pack is at any moment. The number of dogs sounding off at any moment, and the intensity of the baying, gives some indication to the human hunter as to how strong and fresh the scent is. As the dogs signal this information about what they are smelling, the hunter uses it to try to figure out how near the quarry might be. Some control of the pack's movement can then be exerted by humans via signals on a hunting horn, which, to the dogs, simply sounds like a special form of baying.

The sound of baying is also important for the other hounds on the hunt, as well as their human masters. There is a major limitation on the dog's scenting ability, known as *olfactory adaptation*. When you walk into a room, you might notice a faint smell, such as someone's perfume, the scent of flowers in the room, coffee being brewed, or whatever. After only a few moments, you are no longer aware of these smells because of olfactory adaptation. This is actually the result of fatigue in the scent-sensitive cells, which comes about when a particular odor is present in the nose for a period of time. The same thing happens to hounds on the hunt. Typically, when a hound picks up a scent, it will begin to bay or "give tongue." This sound is interpreted by the other dogs in the pack as meaning, "Follow me. I've located our quarry's scent." For a strong scent, however, olfactory adaptation will set in after only about two minutes, and then the hound on the track loses its ability to detect the scent. At this point the dog will go silent and raise its head to breathe fresh, spoor-free air, and allow its nasal receptors to become functional again—a process that will take at least ten seconds and could take up to a minute, depending upon how strong the original scent was. This is why hounds are run in packs. At any given time, some dogs will have the scent and will be sounding off, while others will be mutely running with the pack waiting for their noses to recover. Various members of the pack take turns tracking the scent, so there should never be a moment when all of the dogs are resting their noses at the same time. The dogs whose noses have temporarily shut down know which dogs they should be following, since the dogs still on the trail are the ones that are baying. These sound signals allow the pack to continue to move in a coordinated manner, with every dog still close to the track.

In the same way that whether a dog barks or not is under a high degree of genetic control, so is baying. The geneticist L. F. Whitney noticed that while most bloodhounds bayed when on a tracking a scent, a few rare bloodhounds did not. He was able to demonstrate that by selectively breeding the non-baying dogs, he could produce a strain of silent tracking bloodhounds. Although such a line of dogs might be useful for sneaking up on a hidden criminal, a silent hound would be useless in most circumstances. For instance, since the dog makes no sound, you would not know where it was. This means that you would have to have the dog attached to a tracking leash. Furthermore, since the dog does not bay, you would not know if he had found the scent and was tracking, or whether he was simply strolling through the woods enjoying the normal smells of nature. The baying sound of a dog is thus an important way of communicating what he is smelling to his "nasally challenged" human masters.

17

Dogs Talking to Cats

There is a story, told by my grandmother, Lena, about why dogs seem to hate cats. It is probably of Lithuanian or Latvian origin, as were so many of her stories.

This tale begins shortly after Adam and Eve had been thrown out of the Garden of Eden. It was a magic time, when animals still knew how to talk. They had been given the gift of speech by God, so that each one could whisper their names to Adam, and those names would become a part of human language. In time, however, animals would forget how to talk.

Adam was having great difficulty, since the world outside the Garden was hostile and dangerous. He spent all of his days hunting and farming for sustenance. Even at night, there was no rest. Beasts would come from the forest to try to steal his meager supply of food, or to take his livestock, and even to threaten Adam and his family. He was getting no sleep, and his health and spirits were failing.

Dog had been living in the forest as a wild thing, hunting and scavenging for a living. When Dog saw what was happening to Adam, he thought that here was an opportunity for both of them. He went to the man and offered a bargain:

"I will guard your house at night, so that you can sleep. I will help you to hunt and tend your livestock, so that you can be prosperous. In return, I only ask that you let me rest in your home by the fire, that you give me food, and care for me even when I grow too old to work a full day."

Adam looked at the dog, who was wagging his tail. He knew that when the dog wagged its tail, it was telling the truth and was sincere, so he accepted the arrangement and it became their contract.

It was a fair exchange. Adam could sleep at night and Dog would sound the alarm if the beasts came near, then together they would drive them off. Adam's hunting took less time, since Dog could track and pursue game, and the amount of time needed to tend the flocks was greatly reduced, since Dog did much of the work. As agreed, Adam fed and cared for the dog and gave him a place by the fire.

At the time, Cat was living in the forest as well. Cat was unhappy, because he was basically a lazy creature who preferred to sleep the day away but was forced to hunt for mice in the dense underbrush, or lie in ambush for hours to catch birds. He had seen the home of Adam and it looked good to him. Mice were attracted to the food that Adam stored, and one need only stay nearby, perhaps even in Adam's nice warm house, to catch them. Even better was the fact that Eve threw grain on the ground to attract the songbirds she loved to listen to. Here was an opportunity to have the birds come to him, without long waiting in the cold wet grass. So Cat went to Adam to offer him his contract.

"Man," said Cat, "I will hunt the mice that eat so much of your food and spoil the rest. In return, I ask only that you give me the warmth of your fire, shelter, and some cream or milk now and then."

Adam did not trust the cat, in part because his eyes would close to a slit in the sunlight, and the slitted eyes reminded him of the serpent whose evil dealings had resulted in his family being thrown out of Eden.

"What above Eve's songbirds?" he asked. "I have seen you hunt birds in the forest to kill and eat them."

Cat lied to the man, saying, "I only wish to hunt for the mice and I will leave the birds alone," and then, the clever and devious cat wagged his tail as he had seen the dog do so many times. He could not imitate the dog's tail-wagging perfectly, so when he swung his tail it rippled like a snake, but he knew that tail-wagging was the dog's sign that what he was saying was true and that his intentions were sincere. Adam was fooled by this and accepted the arrangement.

Cat had lied. He did hunt for the mice, but when Adam and Eve were not in sight, he would lurk near where she fed the songbirds, and he would catch and kill them. Eve did not notice because he took the birds to the forest to eat them.

One warm day, Eve was at home and Dog was sleeping in the yard while Adam sheared the sheep in a pen nearby. Cat noticed a songbird near the little pile of grain and killed it. Just as he was about to run and hide with his prey, he heard Eve coming. He dropped the still warm body near the sleeping dog and dashed a few feet away to pretend to be sleeping himself.

When Eve saw the bloody body, she was very angry. "Cat, did you do this?"

The Cat said, "No, it was Dog." Then he wagged his tail in that snakey cat way, and Eve was deceived into thinking that Cat was sincere and telling the truth.

Eve grabbed her broom and began to beat Dog and call him names. She told him that he would have no supper that night and would be tied outside in the cold as a punishment.

Adam heard all of this noise and came running to see what the problem was. When Eve told him what happened, he turned to Dog and asked if it were true.

"I was sleeping and was awakened when Eve hit me. I did not kill the bird, but I have often seen Cat lurking near where the birds are fed." Dog wagged his tail. It was a low and hesitant wagging, but Adam believed that the dog was telling the truth. However, when he questioned the cat, the cat repeated its lies and wagged its tail.

"Both of you seem to be sincere, at least from what I can see in your tails, but one of you must be lying."

"The cat's lies are even in its tail," said the dog. "Look at the wagging of a dog. Our tails are kept straight, like the road between truth and Heaven. Our tails wave like grasses or reeds blown about by God's wind. Yet when Cat wags its tail, it bends and weaves in the image of the serpent who taught the cat to lie."

Adam looked and understood.

"I have misunderstood what I have seen. When a dog wags its tail, it means that he is true and sincere, but when a cat wags its tail, it means that he is planning evil and wishes to deceive you. Dog, whenever you see a cat wagging its tail, you know that he is planning something wicked, so you have my permission to punish him."

Cat protested with a lie, saying, "I spoke the truth," but by now the wagging of his tail when he lied had become a habit. When his tail wagged, the dog immediately ran at the cat and chased it up a tree. That is why dogs have been chasing cats ever since. They watch for the wag of the tail and then know that the cat is up to no good.

My fascination with this story is that it contains an important kernel of truth—not about Adam and Eve, but about dogs and cats. Dogs speak Doggish, cats speak Cattish, and very frequently the same signs and signals can mean exactly the opposite. As I have come to learn more about the language of animals, it seems to me that perhaps some of the enmity and distrust that we see between dogs and cats has to do with their misinterpretation of each other's language.

The very nature of wild felines and canines should predict different language requirements for the two types of animal. Canines live in the social environment of the pack and, as we have learned, communicate in order to define rank and duties, to convey information, coordinate activities, and to minimize conflicts with other pack members. Except for lions, felines are basically solitary hunters in nature. Their interactions with other members of their species often only occur when there are territorial conflicts, or when mating and raising kittens.

Domestic dogs and cats are often forced into close proximity, perhaps sharing the same household, and certainly sharing a neighborhood. Do they share a common form of language? Do they try to communicate with each other? What problems might be encountered when dogs try to interpret Cattish and cats try to interpret Doggish?

Before we can compare the communication signals of cats and dogs, we need to know a bit about cat behavior. As we saw earlier, the dog pack mirrors many aspects of the organization of human society, with a fairly linear, stairway-type arrangement from top to bottom. This keeps the pack structure intact, since when the dominant Alpha dog is not present, the number two dog steps in to take his place and is accepted by the other members of the group as the leader. When cats are forced to live together, they do show a group structure with a dominance hierarchy. However, they interact far less than dogs do. It can't be said that they are asocial or antisocial, but they have not evolved an extensive concept of social organization. The dominant cat acts as if he were on a pedestal, with all of the lower-ranked cats at a subordinate level. This feline "King" will usually be left pretty much alone and be deferred to as far as sleeping space and food. If there has been another cat that has challenged the Alpha, their conflict is not readily resolved and forgotten, and relationships often remain tense and testy.

Subordinate cats do establish a loose ranking system, but it is not linear. Thus, Tabby may be dominant compared to Felix, and Felix may be dominant over Misty, while Misty is dominant over Tabby. Each cat seems to need to challenge each of the other cats to develop a pairwise ranking. Lower-ranking cats do not act like lower-ranking dogs. Instead, they seem to be motivated by defensiveness, hostility, or even avoidance. Rather than showing submissiveness or making pacifying gestures toward a dominant cat, a lower-ranking cat will often simply ignore the existence of the higher-ranking animal in what looks like a sudden case of deafness and blindness to the external world.

For dogs, territory belongs to the pack. Although each dog may have favorite locations, another dog from the same pack or family may occupy that space without causing any conflict. For cats, each individual expresses its status by the amount of territory that it individually holds and is willing to defend. Since cats can also use vertical space, it is usually the case that the top cat not only takes over the largest piece of territory but will also grab the highest and best lookouts for itself, such as the top of the refrigerator or a high bookshelf. This is often the key to peace when cats and dogs must share a space. Since dogs can't climb, resident cats can occupy the elevated areas and feel that their territorial dominance is relatively unchallenged.

When attempting to communicate, cats use sounds, facial expressions, postures, and tail and body movements just as dogs do. Some of the signals are similar to those used by dogs, but many are different. Compare the sounds that cats make to those that dogs use. Cats purr, meow, hiss, spit, growl, scream, chirp, trill, and chitter. Dogs understand some of these sounds, but not others.

The quintessential cat sound is purring. That rumbling sound that we call a purr is clearly meant to be communication, since cats only purr in the company of living things, such as other cats, people, and other pets. All cats purr at the same frequency, which is 25 cycles per second, and this is independent of the sex, age, or breed of cat. The intensity and steadiness of the sound may vary, depending upon the situation. How the cat makes this purring sound is still a mystery. One popular theory suggests that the vibration is set up in a set of "false vocal cords," which are structures near the real vocal cords. Other theories suggest that patterns of contraction of the muscles in the larynx and diaphragm are responsible. There is even one theory which starts with turbulence in the bloodstream, moves to vibrations of the column of air in the windpipe, and ends with sympathetic resonant vibrations in the sinuses. The bottom line is that we don't really know how this sound is made.

We do know that purring starts quite early. Young kittens purr while still nursing. It has been suggested that the mother and the kittens purr as a sign of mutual reassurance. Mother cats often purr when returning to the nest to inform the litter that they are home and all is well. Kittens purr when trying to get other kittens to play. Usually, purring seems to be a signal of well-being and contentment. However, cats may also purr when they are in great pain or suffering from intense fear. It may be that they do this because purring is usually such a positive signal that they gain reassurance from hearing their own purring sounds, much like the behavior of a frightened child who whistles as he walks past a graveyard on a dark moonless night. The familiar cheery noises that he is making give him some encouragement and reassure him that everything will turn out well.

Although other cats may respond to purring, and humans may enjoy this contented sound, dogs seem to be completely oblivious to it. I witnessed an informal demonstration of this when I played a tape of a cat purring in a room that contained four dogs. The dogs were at rest, but not sleeping, when the sound came on. The ears of one of the dogs twitched momentarily, and another stirred slightly and half turned his head in the direction of the sound, but other than that there was no response. The animals did not find the sound of sufficient interest or importance to warrant rising to their feet to investigate or even to make a significant change in body position.

Another sound that is exclusive to cats is the "meow"—a sound uttered with an opened mouth which closes on the "ow" to make a noise that seems to humans to be a distinct word. There are a number of variations in the pitch, intensity, and duration of the meow. There is even the silent meow, where the cat seems to be making the mouth movements associated with the usual sound, but we hear nothing. Actually, analysis of a recording made during one such silent meow showed that there was sound present. Although within the cat's hearing ability, this sound was too high-pitched for the human ear to register.

The cat's meow seems to be a demand for some kind of service. It is made when the cat is hungry and wants to be fed, when the cat finds itself on the wrong side of the door and wants it opened, or just when the cat wants attention. The interesting thing about the meow sound is that it seems to be almost exclusively reserved for human beings. Very young kittens may occasionally make the meow sound to their mother, but after they are grown, cats virtually never meow to other cats or to dogs or other pets. Dogs apparently understand that the meow sound has nothing to do with them, and they seldom seem stirred to move toward a meowing cat.

The only sounds that cats and dogs share in common are snarls and growls. These both are reverberating sounds, but in a snarl the rumbling is accompanied by a curling of the lip to expose the teeth. For both dogs and cats, this sound is designed to increase the distance between themselves and the individual they are addressing. It is an aggressive signal, which is somewhat tinged with fear (the growl being a bit more fearful of the two). Generally speaking, a dog will respond to a cat's snarl by at least stopping its advance, and cats will respond to the snarl of a dog by running away if possible. In that roomful of dogs where we found no response to the purring sound of a cat, playing a tape recording of a cat growling did have an effect. It brought all four dogs to their feet and two of them began pacing around, uttering low-pitched growls of their own.

The facial expressions of cats are somewhat similar to those of dogs. For both cats and dogs, staring, with wide-open eyes, represents a threat. Blinking, for both, is a reassuring signal that breaks the threatening stare. Eyes with half-drooped lids on a dog usually mean contentment and relaxation, while in the cat they represent trust and relaxation. Thus, the meanings of eye signals are similar enough so that there is little chance that much misunderstanding will occur between the species.

The ear signals of dogs and cats are also somewhat similar, as can be seen from Figure 17-1. As with dogs, a happy and relaxed cat will have its ears up and facing forward. A calm, attentive cat will have its ears up, but angled slightly forward in the direction of the interesting event. Again as in the case of dogs, a frightened cat will tend to flatten its ears down against its head, although sometimes they will flatten out to the side, giving the impression of airplane wings.

Aggression is the ear message that is different in cats and dogs. As we saw earlier, a dominant aggressive dog will hold its ears erect and forward. Just before initiating an attack, the ears might tilt slightly out to

RELAXED

ATTENTIVE

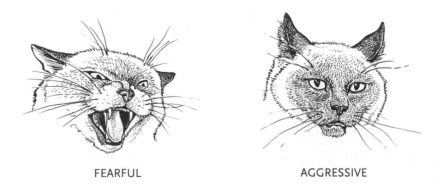

FEARFUL

AGGRESSIVE

Figure 17-1   The basic ear signals and facial expressions of a cat.

the side, widening the apparent V-shape that the upright ears make. For the cat, this gesture is much more pronounced, and the ears turn, so that the inside surfaces are directed down toward the side, and the backs of the ears are now visible. In some of the larger wild felines, the backs of their dark ears are actually marked with pale patterns, which makes this rotation of their ears more conspicuous when they are signaling their aggressive intentions. Other wild cats have tufts at the ends of their ears, which also help to make the ears' positions and rotations more visible from a distance.

Once we reach the tail, however, just as in my grandmother's folk story, chances of misunderstandings between cats and dogs become much greater. Although dogs and cats wag their tails, their meanings are actually opposite. For dogs, the broad wagging of its tail is what we might call a distance-reducing signal, which invites another individual to come close in a friendly manner. For cats, however, tail-swinging is a distance-increasing signal, which tells the viewer to go away and signals some form of emotional conflict or tension. Cats often begin to flick the tip of their tails back and forth before pouncing or slashing with a paw. The broad swinging motion, which may actually thump the floor as it increases in speed and arc size, is the clear signal of aggression in a cat. If there are no other aggressive signals visible, a dog might be misled into approaching this "tail-wagging" cat in a friendly manner, only to be met by claws and teeth. Thus, it would not be surprising to have the dog view the cat as a "liar," unworthy of trust on future occasions.

More static tail positions can also deceive. Both dogs and cats will tuck their tails close to their bodies and lower themselves down to make them look smaller to signal abject fear and submission. However, the Doggish and Cattish meanings begin to diverge for other tail positions. Dogs who are showing a fair level of submission and acceptance of lower status will drop their tail down and let it hang straight down over their rump with almost no movement. Cats have a similar position in which the tail hangs down to form a sort of inverted letter L-shape, shown in Figure 17-2. However, in the cat, this vertically hanging tail is not a submissive gesture but rather a signal from an animal who is taking the role of aggressor. The aggressiveness of this signal is even greater if it is accompanied by a slight arching of the back. Thus, a dog who sees a cat give the down-hanging tail signal that in Doggish means that it is not a threat will most certainly be unprepared for the cat's aggressive rush. Alternatively, the cat may misread the dog's downward-hanging tail signal, which was meant to convey that it wants a truce, and interpret it instead as the beginning of an attack.

FRIENDLY GREETING   AGGRESSIVE

DEFENSIVE   FEARFUL

Figure 17-2   Typical tail positions and body language of a cat.

For dogs, the tail held straight up, or curved somewhat over its back, is a sign of confident dominance, and an assertion of authority. For cats, this same signal is one of the friendliest signs the species ever offers. The high tail, with perhaps a slight forward curve over the back, allows another friendly cat to investigate the exposed region under the tail. Cats, just like dogs, have scent glands around their anal region which provide pheromones that can identify the individual to anyone who catches the familiar odor. Raising the tail is much like offering someone your passport or driver's license to verify your identity. A dog that sees this upturned tail signal from a cat is apt to misinterpret this Cattish message of friendship as an attempt to express dominance. The cat may similarly feel betrayed when its offer of friendship is met with suspicion and threat.

A vertical tail position can also have another meaning in cats. The frightened cat bristles its fur, arches its back, holds its fluffed-out tail erect and vertical. This posture, which produces the typical Halloween cat silhouette, means that the cat is very frightened about the current situation. In dogs, *piloerection* (the technical term for the bristling or fluffing up of an animal's fur) is designed to make them look larger and represents a high level of aggression. As we saw earlier, when a dog is communicating a potential attack, we usually see the fur rise on the shoulders (the hackles), and this may extend in a ridge down the center of the back. At the same time, the tail may bristle out as the dog raises it into a dominant vertical position. If a cat sees this display in a dog, it may misread this as the Cattish statement suggesting that the animal it is looking at is terrified. Then, by failing to back down, the cat may actually trigger an attack. The dog can just as easily misinterpret the cat's indication of fear. If it reads this signal in Doggish, the bristling fur and a raised tail mean that the cat is not backing down and is insisting upon a fight.

Body language is also different for cats and dogs. Take the language of submission in a dog. A fearful cat, like a fearful dog, will lower itself to the floor to make itself appear smaller and signal that it is no threat. A dog that is terrified will go even further, rolling on its back to expose its unprotected stomach. This is the ultimate sign of submission in a dog. A cat, however, does not roll onto its back in fear or submission but rather as a defensive movement or when it is trying to kill some prey. This position puts it into an attack position by freeing all four sets of claws. A cat who is hunting a large or moderate-sized animal, such as a bird or rat, will usually pounce and then quickly roll onto its back. It now grapples with its prey by clutching at it with its front paws and biting. Meanwhile, its rear paws slip under the animal's stomach area, the claws unsheathe, and the cat kicks out with both hind paws simultaneously. This movement can disembowel or greatly damage the vital organs of the cat's victim or an aggressor. The opportunities for canine and feline misinterpretation are obvious. An angry cat rolls onto its back, which the dog reads as the Doggish message that the feline is giving up the fight and now wants peace. He approaches the cat to give the ritualized sniff, which is the Doggish acceptance of such an offer of truce, only to have his face raked by four sets of flashing claws.

Misunderstanding between the species can also occur in more subtle signals. Remember that dogs raise a paw when they are tense, moderately fearful, or trying to attract attention from someone they view as more dominant. Remember that this gesture became ritualized, since it is the first movement required if the dog is going to roll over on its back to show submission. For a cat, this paw motion also precedes rolling over on its back, but here it is a preparation for aggression. At low levels of threat, the cat will raise one paw and gesture gently in the direction of threat or an annoyance. If a dog misinterprets this gesture, he may approach the cat, which might only provoke having his face or nose swiped by the cat's claws.

There is one last area where dogs and cats have a major opportunity to misread one another, and this has to do with direct body contact. We saw earlier that dogs often bump against people or other dogs, or stand against them and lean their weight as a dominance signal. For cats, however, rubbing their bodies against something, using shoulder, chest, or head bumps, or leaning heavily with their hips, is a means of leaving their scent behind on that object or individual. This mixing of scents helps to identify familiar from unfamiliar and friend from stranger, so it is part of a cat's friendly greeting ritual. If a cat does this to a dog, it is easy for the dog to misread the Cattish welcome gesture as an attempt to show dominance by leaning against him.

Obviously, many opportunities exist for misinterpretation of signals between cats and dogs. It is tempting to speculate that much of the dislike cats and dogs have for one another may simply be due to the fact that Cattish and Doggish languages have opposite meanings for the some of the same signals. An animal that suffers injury, fear, or discomfort because it has misread a signal from the other species is not likely to enter its next encounter with a feeling of trust. Starting the encounter with a sense of hostility can easily escalate into the classic conflict we observe between dogs and cats.

It is possible for dogs and cats to become "bilingual." In the same way that dogs, and to a more limited extent cats, can learn some aspects of human language, they can also learn to read each other's signals if they live in the same home. A puppy and a kitten who are reared together seem to have few problems, mostly because their misunderstandings are worked out when both are young. As we saw earlier, a puppy who shoves his nose into a kitten's belly when it has rolled over onto its back will be taught the consequences. However, at this stage of development, the tiny claws and teeth provide little chance for damage, and lots of opportunities for learning.

Bringing together adult cats and dogs is more difficult, especially if the dog has lived with or associated with other dogs, and if the cat has socialized extensively with other cats, since animals with such prior experience will have developed strong expectations as to the behaviors which their Doggish or Cattish signals should lead to. Probably the best course of action is simply to let the animals work out their language difficulties. Experiencing the consequences of misinterpretation usually produces the fastest learning, although at the early stages of introducing a dog and cat into the same home, they should be monitored. If things get out of hand, and the aggression level rises to the point where one or the other seems to be sustaining physical damage, it is best to intervene.

Don't rush in to try to be a cross-species interpreter or referee when trouble breaks out between your cat and dog. All you will get for your efforts is scratch marks from your cat and bite marks from your dog. The simplest thing to do is distract the combatants. Do this from a long distance, by squirting some water at them from a water pistol or a plant sprayer, or even emptying a glass of water over them. Another strategy is to throw a blanket, bath towel, or coat over the dueling animals. When they break off their confrontation, remove one animal (preferably the one who appeared to be on the losing side in the confrontation). When both animals are calm again, perhaps after an hour or so has passed, allow them both access to the same areas again. The language lessons will continue for a while, but soon enough some harmony should be achieved.

The best signs of how the animals are acclimating to each other can be found in their sleeping arrangements. Until the cat and dog get used to each other, they will seek out separate sleeping areas, such as separate rooms where they are effectively hidden from each other. As they develop mutual trust, they will begin to sleep in the same room. The greatest sign of trust, and a good signal that they have reached some permanent accord, is when the animals are willing to sleep in the same room, near each other, with their backs turned toward one another. After all, you simply don't turn your back on someone that you don't trust.

Dogs and cats that accept one another as part of their mutual pack or living group may ultimately engage in a common form of friendly communication: grooming one another. On most mornings, when my alarm goes off, our orange cat Loki jumps down from his sleeping area on the window seat. If there is a bit of sunshine, he wanders over to the glass doors to our sundeck and lies down in front of them. At about the same time, my flat coat retriever, Odin, gets up from the pillow next to the bed where he sleeps. The big black dog stretches, yawns, then wanders over to the cat and proceeds to lick his face and body, grooming him much as a mother might groom her puppies. He then lies down next to Loki, and the cat conscientiously cleans his ears and face with his tongue. Odin and Loki have been together since the dog was nine weeks of age and the cat eight weeks. Having grown up together, they clearly have worked out their linguistic misunderstandings, so that Odin speaks a bit of Cattish and Loki speaks enough Doggish to get by. Of course, if my grandmother's tale is correct, my dog still occasionally checks our cat's tail just to see if he is lying.

18

Doggish Dialects

Several aspects of language are nearly universal in all animal languages, or at least in the language of mammals. However, our consideration of the differences between the communication patterns of cats and dogs should have made it clear that all animals do not use exactly the same signals with exactly the same meanings. It is interesting that even if we restrict our consideration to the language of dogs, we will find systematic differences between groups of dogs—what might be called "dialects."

Our current domestic dogs differ from wild canines, such as wolves, in a number of ways. The most important of these involves *neoteny*, which is the technical term describing the fact that certain juvenile features and behaviors are retained in the adult. This means that an adult domestic dog is shaped much more like a puppy than is an adult wolf, with a shorter muzzle, a wider and rounder head, somewhat smaller teeth, and it may also have floppy ears. Behaviorally, dogs are also more like wild canine puppies, with their lifelong desire for play. Furthermore, as we found out earlier, barking is not a behavior normally seen in adult wolves, but is a characteristic of wolf puppies, and of course, adult domestic dogs. In effect, our dogs are the Peter Pans of the canine world.

Domestication and neoteny seem to go hand-in-hand. During the early stages of human contact with dogs, before people were actively trying to breed dogs as companions and workmates, dogs appear to have begun to domesticate themselves. The evolutionary principle of "survival of the fittest" works regardless of the environment in which it is applied. The "fittest" canines, in the case of these early opportunistic scavengers, would be the friendliest and least threatening, since they were most likely to be allowed to get closer to the human campsite and thus get at any available food resources, such as leftovers and garbage. This "evolutionary pressure" to be friendly became greater when humans began to actively try to domesticate dogs by controlling their breeding. Obviously, a canine that was vicious or fearful of people would not fit in well in village life. Such unsociable creatures would be driven out of the settlement or simply killed. Dogs that were friendly toward people would be much more trainable and hence more useful. These were the canines that would be kept and nurtured, and so also the dogs that would go on to parent the next generation of dogs. However, this process has some unexpected side effects.

In the late 1950s, the Russian geneticist Dmitry K. Belyaev began a project which has so far been going on for over forty years.[1] He theorized that nearly all of the physical and behavioral differences between domestic dogs and wild canines has resulted from the simple act of selecting canines for friendliness toward humans or tameability.

Actual experimental studies on the process of evolution are difficult to set up and conduct. However, working at the Siberian Department of the Russian Academy of Sciences in Novosibirsk, Belyaev decided to try to turn the clock back to the point where active efforts to domesticate dogs began. He could then "replay the process" and carefully study what was happening during the creation of the dog. When it came to the actual selection of which wild canine to use as his "protodog," he decided not to use wolves. This was a considered scientific decision, since wild wolf stocks are no longer genetically "clean." It is well known that domestic dogs have often escaped and bred with wild wolves, so that would make any interpretations more difficult. Instead, he chose a canine species which is very close to dogs, but will not naturally breed with them, and which has never before been domesticated: the Russian silver fox (*Vulpes vulpes*).

The actual experiment was conceptually very simple, but involved a lot of work and patience. Beginning with 130 undomesticated foxes, he established a systematic breeding program. Each new litter of foxes were tested for friendliness toward humans. In the early generations, to be selected for further breeding, the foxes had to be willing to allow themselves to be hand-fed and stroked by humans. This quality was found in only around 5 percent of the early generations. Later on, Belyaev and his associates began to make the requirements more stringent. To be kept as breeders by the sixth generation, the foxes also had to actively seek out human contact, by approaching with their tail wagging, and had to seek human attention by whining. In each successive generation, only the most tame and friendly foxes were kept. Thus, it was apparent that successive generations were becoming more and more like domestic dogs in their behavior. The animals bred for friendliness would approach people, then lick and sniff at them to try to elicit affectionate responses and petting.

There have now been more than 35 generations of breeding, involving some 45,000 foxes over the four decades that this project has run. Soon the scientists found they had a surplus of these "domesticated foxes." At the same time, they were faced with decreased funding due to a weak Russian economy. The solution to both problems was to sell off the surplus animals as pets and to use the resulting funds to keep the research project going. They did follow the progress of some of these animals to see how they were doing in their new homes, and they found that when adopted into typical human households, these domestic foxes did well. The owners described them as being "good-tempered" companions and pleasant pets. They bond well with people, although they are a bit more independent or catlike than most dog breeds.

One of the important outcomes of the research is that although the foxes were being selected on the basis of only one behavioral characteristic, namely, friendliness, they began to change physically. Floppy ears, curled tails, shorter tails, lighter-colored or even multicolored coats began to appear; in addition, their muzzles grew shorter, their heads a bit rounder and broader, and their teeth somewhat shorter. All of these transformations are similar to the changes that distinguish domestic from wild canines. The whole time course of development, from puppy to adult, was altered during the process of selection. In foxes as in all canines, there is a systematic sequence and a relatively fixed timing when certain puppy behaviors appear and then are left behind. When these are measured, it becomes clear that the timing and rate of development has been extended by the process of domestication. Puppylike behaviors appear very early and linger on much longer in these domesticated foxes than in the wild fox. In other words, not only to we now have domestic foxes; we have foxes who, like dogs, maintain many juvenile features and behaviors into adulthood. Thus, Belyaev's work clearly shows us what has really happened during the process of domesticating dogs: breeding for tameability and friendliness has resulted in dogs that are both mentally and physically more like wolf puppies than wolf adults.

Although all of Belyaev's selections were made on the basis of friendliness, early humans may have also been selecting for that puppy "look" as well during their initial efforts at dog breeding. It is an obvious fact that animals and humans alike instinctively feel some special fondness for the young of their own species. Naturalists, such as the Nobel Prize winner Konrad Lorenz, have suggested that this feeling may be triggered by something about the appearance of young animals. In essence, they seem "cute" because they are small and have big eyes, round, flat faces, and appealing facial expressions, and make high-pitched sounds. It turns out that this "cuteness" is actually a survival factor, making adults more protective and solicitous of the younger animals in their group. Contemporary psychologists have shown that this cuteness factor crosses species boundaries as well. We tend to feel more warmly toward kittens than adult cats, and chicks appear more attractive to us than adult chickens. The same goes for puppies as compared to adult dogs. It is difficult not to want to take virtually every puppy that you meet home with you. Early man, and perhaps to a greater degree early woman, would probably have thought that the more puppylike of their newly domesticated dogs were cute, just as we do. Perhaps the cutest got the best care. Perhaps they were fed first and got the bone with most meat. Perhaps they were the ones selected to share human shelter, protected for inclement weather, and given more opportunities to breed.

Domestication has not only affected the looks and general behaviors of dogs; it also may have altered the contents of our dogs' vocabulary in comparison to that of their wild cousins. One might say that in domestic dogs, the ancestral social behavior and communication patterns of the wolf are fragmented and incomplete. The dogs' behaviors form a sort of mosaic that contains some of the adult wolf communication signals, but also contains many juvenile signals.

If we look at the language development of wolves and wolflike dogs, we can see a developmental sequence for the appearance of certain communication signals. Very young puppies are helpless and dependent, and therefore most of their signals are oriented toward soliciting care and showing submissive, yielding, and pacifying behaviors around adults. Thus, the young puppy is more likely to lick the muzzle of an adult, or crouch low, or look away. As the dog grows older, other more socially dominant signals begin to appear in its vocabulary. The adult dog is more likely to give a threatening stare, to growl, or to stand over another dog. You can actually plot the appearance of these signals over a dog's lifetime. The simpler submissive signals tend to appear early, and the dominant and more socially complex submissive signals later, as the animal becomes an adult. If we call the adult language "Wolfish," and the juvenile language "Puppyish," it should be clear that a speaker of Wolfish should still be able to understand Puppyish, because he spoke it when he was younger. An individual who speaks only Puppyish, however, is at a disadvantage, since he may not yet have learned all of the terms in Wolfish. This is the problem that domestic dogs may have relative to wolves. They speak Puppyish. They may have some receptive language ability in Wolfish, but their productive vocabulary is limited because neoteny has stalled them before they achieved the full set of adult linguistic capabilities. This has been known to make the communication between domestic dogs and wolves difficult. In one study, Malamutes were raised with wolves, and they often failed to correctly read the social behavior signals of the wild canines.

Now let's complicate things just a bit. All domestic dogs do not show the same degree of neoteny. Probably the best indicator of how much or how little neoteny a breed has is how similar the individuals are in their appearance to an adult wolf. Dogs that look a lot like wolves, such as German shepherd dogs and Siberian huskies, not only have more adult physical characteristics but show less neoteny in their behaviors. Conversely, dogs that look more like puppies, such as the Cavalier King Charles spaniel or the French bulldog, should not only have more juvenile characteristics physically but also behaviorally because of their greater neoteny.

It takes only one step from these observations to conclude that different breeds of dogs may develop different dialects or versions of the Doggish language. Those closest to the adult wolves in their characteristics are most likely to use many elements of Wolfish in their language, while those with a greater degree of neoteny may be relatively illiterate in Wolfish and speak only Puppyish versions of the canine tongue. Deborah Goodwin, John Bradshaw, and Stephen Wickens, who are researchers from the Anthrozoology Institute at the University of Southampton in Great Britain,[2] studied ten different dog breeds, which they rank-ordered in terms of their similarity to the wolf. From most puppylike to most adult wolflike, their list was:

1. Cavalier King Charles spaniel
2. Norfolk terrier
3. French bulldog
4. Shetland sheepdog
5. Cocker spaniel
6. Munsterlander
7. Labrador retriever
8. German shepherd
9. Golden retriever
10. Siberian husky

The researchers then looked at fifteen different dominance and submission signals. What they found was quite consistent with the notion of neoteny, not only for body shape but for dog language. The least wolflike of the dogs, the Cavalier King Charles spaniel, had the most limited social vocabulary, consistently showing only two of the fifteen targeted social signals. These two signals were also the earliest to appear in the normal development of a wolf, and would be shown by a three- or four-week-old puppy. It seems as though the social vocabulary of this breed stalls at this level. The Siberian husky, however, shows the full display of tested social communication signals, thus giving a behavioral vocabulary similar to that of an adult wolf. For the breeds between these two extremes, the more wolflike they were, the greater the number of social signals they will use, and the more likely the gestures employed will be those that develop later in life.

It is important to note that this research is talking about dog communication, not personality. These results do not mean that a Siberian husky, golden retriever, or German shepherd is necessarily going to be more aggressive than other breeds. Instead, the results mean that these less neotenized breeds have a greater vocabulary of signals and gestures that they can use to communicate with other dogs about social matters. They not only have a greater range of aggressive signals, but also a greater range of pacifying signals. If one remembers that part of the purpose of dog communication is to cement good social relations in the pack and to avoid physical confrontations where both parties might get injured, these more wolflike dogs have the greater response range, and perhaps the greatest ability to have a subtle "conversation" about social status. Ultimately, this might result in a greater ability to avoid direct conflict. Those breeds that speak a Puppyish dialect will have a more limited vocabulary and usually a greater range of submissive, rather than aggressive signals. They may also be less aware of signals given by other dogs to indicate social ambitions, assertions of rank, or even surrender to a more dominant dog.

One obvious effect of these linguistic differences is that there may be misunderstandings between dogs that speak different dialects. Important signals may be overlooked by the more puppylike animal, whose dialect does not focus strongly on social dominance. The less linguistically skilled animal might inadvertently trigger an actual physical attack, or the conflict might be continued beyond the point of surrender because the dog who speaks the more adult Wolfish dialect was looking for a particular signal to indicate submission, and not finding it, proceeded to escalate the level of aggression.

Let's consider one such case, which was described by the Nobel Prizewinning novelist John Steinbeck, who wrote such classics as *The Grapes of Wrath, East of Eden,* and *Of Mice and Men.* Steinbeck had a love of dogs and one of his books, *Travels with Charley,* describes a nearly year-long trip that he took with his black standard poodle, Charley, as his only companion. This particular case, however, involves a dog that Steinbeck owned much earlier, an Airedale. Based on their looks, Airedales do not appear to be very wolflike. Steinbeck describes an ongoing territorial dispute between his dog and another. This second dog appeared to be much more wolflike in its looks, and the author describes it as a "shepherd-setter-coyote mix." It seems that whenever his Airedale went past the domain of this particular dog, a fight would break out. Steinbeck reports: "Every week my dog fought this grisly creature and every week, he got licked."

This lopsided war went on for several months. Then, one day, Steinbeck's Airedale got lucky. He caught the tough, wolflike mongrel by surprise and really gave him a thrashing. Steinbeck appears to be very saddened by what happened next. The beaten dog "hung his head in the loser's corner." In typical passive and submissive manner, he rolled on his back and exposed his vulnerable underside. At this moment, according to Steinbeck, the Airedale "abandoned all chivalry." To the author's dismay, the Airedale wasn't satisfied with just winning the fight. Instead, while the losing dog lay on its back, and in true wolflike manner signaled its submission to end the conflict, Steinbeck's dog suddenly returned and savagely attacked his private parts. The scene was quite horrible. By the time the Airedale was pulled way from his victim, the latter "was finished as a father." Steinbeck ends his account by saying, "There can be dogs without honor, even as with us."

Steinbeck may be correct in assuming that this particular Airedale was simply malevolent or demented, and deliberately wanted to inflict pain on his former tormentor despite his obvious surrender. However, another possibility is that being from a breed which is much further from the wolflike prototype, he simply did not fully understand the social significance of the other dog's gesture. For a Wolfish-speaking dog, this submission signal would mean a permanent decrease in rank. Socially, then, this signal is important, and represents a critical message. For a dog who speaks a less Wolfish dialect, such social signals may simply not be as easily read, or they may be interpreted as having only a momentary importance, without the long-term nuance of submission which the Airedale needed to feel safe. Failure to recognize the signal and its full meaning could explain why the Airedale did not end the attack when he saw the message of defeat from the other dog. If this is true, then it was ignorance of the language, rather than evil intent, that caused this violation of canine tradition and code of honor.

Of course it is not true that all dogs will respond appropriately and predictably to a communication any more than it is true that all human beings will respond appropriately and predictably to a spoken message. There are breed variations in the dialects spoken, and there will also be individual differences between various dogs in their responses. This was running through my head as I watched an extremely tense situation develop in a dog obedience class that I was visiting. It was the first class of a new session, and a woman had brought in the largest German shepherd dog I have ever seen. The dog was called Shredder; this seemed appropriate, since he was giving a full set of aggressive threat signals to all the other dogs that came close. Furthermore, any human who came too close was treated in the same way. The other new students had all backed away and seemed to be pushing themselves against the farthest wall while protectively clinging to their own dogs. The instructor, Ralph, recognized he had a problem.

"Is he always like this?" he asked.

"Only when he gets tense," the woman answered in a quavering voice.

"Well then, I'll just have to tell him, in dog language, that he's not being threatened," said Ralph.

As Ralph reached into his pocket to get a few treats, I thought I knew how he would handle the situation. Instead, much to my amazement, he lowered himself to the floor and sat with his legs spread in the direction of Shredder.

"This is the human equivalent of a dog's submissive posture," he explained. "As far as he is concerned, I am exposing my underbelly and genitals, which means that I'm not a threat to him. No dog will ever attack another dog in this position."

I held my breath while Shredder approached Ralph's widespread legs, still growling. The instructor placed his hand between his legs (a bit defensively it appeared to me) and opened his hand to reveal a treat. Shredder cautiously continued his approach, and then slowly took the treat. He sniffed Ralph's crotch, then turned his side toward the seated man. When Shredder finally turned back and sat looking at Ralph, the man carefully stood up.

"Okay now, just take a seat over there, and I'll deal with him separately," he said.

"Wasn't that a little foolhardy?" I asked.

"No, it was perfectly safe if you know dog language."

I thought of John Steinbeck's Airedale. When he was spoken to in Doggish, he certainly did not react as expected. It may be that he didn't know that particular signal in that dialect of Doggish. Alternatively, he may have known dog language and simply chosen to ignore the message. The fact that we know how to communicate does not guarantee that the dog will be willing to listen and respond to what we are saying. It depends upon the breed, the individual dog, and the situation. This time Ralph got away with his maneuver: the dog read his signal and responded appropriately. The German shepherd has a lot of similarity to timber wolves, and does not show extreme neoteny, so it might well be sensitive to this kind of message. However, next time, Ralph might choose to try this on the wrong dog—perhaps one who speaks Wolfish less fluently and has less of a sense of dominant and submissive signals. I shudder to think of the possibilities.

19

Is It Language?

Early in this book, I told you that I was going to use the words "language" and "communication" interchangeably and not worry particularly about the scientific debate about the difference between these concepts until later. Now that we know more about Doggish, we can take a look at this question. It is necessary to explore the issue because the scientific controversy is very real, and I am, by profession, a scientist. So let us see if dogs do have language as we humans understand the word, or whether their communication is only a collection of signs and signals.

In most areas of science, when you use the word "language," it is defined as a method of communication that uses utterances, signs, symbols, or gestures to convey meaning. However, within this broader definition, there are also some specific requirements. Historically, the list of these requirements was long and was drawn up in such a specific way that we would end up concluding that only human beings have language. Today, the list is much shorter, perhaps because we recently became more comfortable seeing humans as part of the natural world, and not an exclusive or special creation resting on an evolutionary pedestal. Most psychologists and linguists today would probably agree on around four or five basic requirements for something to be defined as a language.

The single most important feature of a language is *meaningfulness* (sometimes termed *semanticity*). This one is obvious, since the only purpose that language serves is to communicate meanings to others. Words must refer to things, ideas, actions, or feelings. While individual words must have a meaning, specific combinations of words can also modify or clarify those meanings. I think that we have established that Doggish signals clearly do have meaning. Dogs do not bark, growl, raise their tails, or stare at you randomly to no purpose. In fact, at the end of this book I have supplied a "Phrasebook" (see p. 253), which provides a sort of dictionary of the meanings of Doggish signs and symbols, so we can presume that this requirement has been met by canine communication.

The next basic requirement is *displacement,* which refers to the fact that language permits communication about objects and events that are "displaced" in space or time. This simply means that you can use language to communicate about objects that are not present and visible at the moment, or events that have occurred in the past or may occur in the future. Although, at the productive level, dogs usually don't discuss absent objects very often, their ability to understand linguistic constructions that involve them is clear. Most dog owners have a variety of "Find the object" phrases that they use. For example, my dogs respond to "Where's your ball?" by dashing around to find the ball and then bringing it to me. If it is inaccessible, they will usually stand near it and bark. "Where's your stick?" causes a search for the last stick that the dog was playing with. "Where's Joannie?" is a convenient phrase which helps me to locate my wife. Upon hearing it, the dog goes to the room where my wife is. If she is upstairs or in the basement, the dog will move to the appropriate set of stairs and wait there. If she is out of the house, the dog will usually go to the door she used to leave. If the dog doesn't know where she is, he will usually start to search for her. In all of these cases, the dog is responding appropriately to an object that is not immediately present, which fulfills the requirement of displacement. As for evidence of displacement in productive language, there is less; but remember that dogs will give a specific "Call the pack" alarm bark, even though the other pack members are out of sight at the moment.

When it comes to the question of whether dogs have "true language," in the sense that humans have language, one of the great stumbling points has been the issue of grammar. Grammar is the set of rules by which we structure a language. Among the most important part of these rules is syntax, which deals with the order in which words and phrases are put together. For example, in English, the article "the" goes in front of the item that it refers to. Thus the sentence, "The boy threw the ball," makes sense, while the sentence, "Boy the threw ball the," where the position of "the" is twice reversed, makes no sense in English. The specific rules that determine the order by which we combine different parts of speech can differ in different languages. In English, the adjectives that describe an object are usually in front of that object, as in the phrase "white house." In French and Spanish, that order is reversed, and we would say *maison blanche* or *casa blanca.* The rules that determine the order in which parts of a language can be combined also determine which words can meaningfully go together. Phrases like "these cat" or "an ball" are not sensible in English. We can call this aspect of grammar the *rules of combination.*

The specific order of the words can also determine the meaning of what is said. For example, the phrase "man-eating shark" refers to something quite different from the phrase "shark-eating man." Similarly, "The boy hit the girl," means something quite different from "The girl hit the boy." We can call this aspect of grammar *rules of word sequence.*

Do dogs have grammar in these two senses, rules of combination and rules of word sequence? Until recently, most scientists seemed to think that the answer to this was no. However, based upon some recent observations, there are some tantalizing suggestions that dogs may show at least some evidence of having grammar.

Consider the rules of combination, which allow some things to go together in language, and bar other possible combinations. When we consider the sounds which dogs and wolves make, we find that there are some combinations that never occur together. Howls and whimpers are an unheard-of combination. You will also never hear howls and growls together. On the other hand, howls are happily combined with yips, and occasionally with some types of barks. Barks can be combined with other barks, with growls, and with whimpers, but growls and whimpers are never combined with each other.

Since so much of a dog's language involves posture and body signals, it is interesting that certain sounds are never combined with certain body postures. One never sees a stiff-legged dominant stance combined with either whimpers or yips. This stance is usually produced along with growls and, less frequently, with an alerting bark. The rollover position where the dog exposes its stomach submissively is never combined with growls or barks, but may be accompanied by whimpers or whines. The raised-paw, uncertainty position also is never combined with growls or barks, and, in fact, seems most frequently to be a completely silent gesture.

We can also find tail gestures that follow rules of combination with sounds. The high-curled tail of the confident dog is never seen in conjunction with whines, whimpers, or even growls. Before a confident dog will begin to growl, he will uncurl his tail, so that it is straight and in an upward angle, pointing backward. When this "let's see who's boss around here" tail signal is displayed, whimpers, whines, and howls are never heard.

In fact, there are many body, tail, ear, and mouth expressions that often have certain vocal accompaniments and seem to never be combined with other vocal sounds. All of this, taken together, certainly seems to suggest that dogs have some elements of grammar associated with the rules of combination.

Perhaps the most exciting recent observations are those suggesting that dogs may also have grammar in the form of the rules of word sequence. Let us consider two simple sounds dogs make. The first is the curled-lip growl, which sounds something like "harrrr." Taken by itself, this growl is a confident warning for another dog or a person to stay away. It may be heard in situations where a dog has prized object, like a nice bone, or a bowl of food, where this growl is used to mean, "Back off—this is mine!"

The second simple sound is a bark, which starts low, rises in pitch, and ends with something like an "f" sound. It can crudely be described as "rrruff." This is the common alerting bark that dogs give to get the attention of other members of the pack. "You might want to come over and look at this." It is normally responded to by other dogs moving in that direction to stand near the one who barked.

When we combine these sounds, however, we get different meanings, and the specific meaning depends upon the order in which they are combined. The combination "harrr-rrruff" is actually an invitation to play, and is usually combined with the typical play-bow. Reversing the combination, to produce "rrruff-harrrr," results in quite a different message. It is a threat uttered by an insecure dog, perhaps trying to protect an item like a bone, but sometimes just to ward off another dog who may appear to be dominant and threatening. In this form, the sound means something like "You're making me nervous, and if you come any closer, I may be forced to fight." The fact that this signals a threat based upon insecurity is what makes it different from the simply "harrr" sounded by a secure and dominant animal.

As human beings, we are biased toward viewing everything in terms of our own language. So we tend to look for grammar combinations and sequences of words in the form of sounds. If we take the viewpoint of the dog, however, for whom a body signal is just as important as a sound, it may be possible to find other evidence for rules of word sequence. When one dog stares directly at the front of another dog's face, this is generally a dominance display or a threat, which basically says, "I think I'm top dog here. Do you want to challenge me?" On the other hand, a dog that deliberately breaks off eye contact with another dog and looks off to the side is showing that it is not a threat and is basically saying, "I accept the fact that you're the top dog around here. You can set the rules and I'll do what you want." Combining the two signals, so that it begins with a full-faced stare, followed by a brief looking away and then looking back, changes the meaning, to suggest a more peaceful meeting between two dominant dogs. This can be interpreted as something like "You're certainly tough and may be the boss around here. Well, I'm pretty tough, too, but let's not fight."

Now, let's take these two signals and combine them with a sound. In so doing, we can completely change the nature of the communication. When a dog gives a full-face stare at another, and at the same time gives the curled-lip growl "harrr," the likelihood of physical confrontation is very high. This is the canine equivalent of the traditional western showdown where the outlaw in the black hat announces, "This town is not big enough for both of us. Draw your gun." However, if a dog gives another dog a full-face stare and then looks away and utters the growl "harrr," the response is quite different. The dog who was stared at now looks in the direction that the growling dog is looking. He may also adopt a defensive posture beside the other dog, looking in the same direction. This exchange means something like "I think there's trouble over there. Let's team up and take action if it's necessary."

The important thing about these exchanges is that a particular element, whether it is a sound ("harrrr" or "rrruff") or a body language gesture (a full-face stare or averting the eyes and the head), takes its meaning from where it is placed in a sequence of sounds or gestures. This certainly seems to suggest that dogs do use grammatical rules of word sequences.

Taken together, observations like these indicate that the language of dogs is more complex then we have previously thought. There certainly seems to be some evidence for at least a rudimentary grammar and syntax, with evidence for both rules of combination and rules of word sequences.

The last basic requirement for language is known as *productivity*. A true language must allow the expression and comprehension of an infinite number of novel expressions, all of which are created on the spot. More clearly, this notion is based on the assumption that language is a creative system of communication, as opposed to a repetitive system that works on the basis of recycling a limited set of sentences or phrases. Some researchers might suggest that this requirement seems to rule out canine language. Unfortunately, interpreted strictly, it would also rule out any simple language that has a small vocabulary and limited grammatical rules which keep the sentences short. A child of two or three years of age, with a vocabulary of only one hundred words and a sentence length limited to two words, will have a fixed number of possible sentences and will "recycle" these sentences as needed to communicate with those around him. Yet we give this child credit for having language, even though it fails the test of productivity.

My predisposition is to accept Doggish as a simple language, using the same rules and criteria that we use to credit young children with language. When testing language development in people, in addition to sounds, psychologists do recognize gestures as language components. Consider one such test, the *MacArthur Communicative Development Inventory*, which has forms for measuring language development for children as young as two years of age. It has an entire section on "Communicative Gestures," which it counts as language. These include pointing to interesting objects or events, waving "bye-bye" when a person leaves, extending the arms upward to signal a wish to be picked up, and even smacking the lips in a "yum-yum" gesture to show that something tastes good. Certainly, the communicative gestures of dogs are equal in complexity to these.

In drawing the similarity between canine communication abilities and infant speech we must not go overboard. There are unavoidable parallels, however. In both dogs and human infants, receptive vocabulary is larger and more reliable than productive vocabulary. The linguistic items that are understood are also more likely to contain information about actions that the speaker would like the child to perform. We say, "Give me your hand," to a child and grant it some linguistic ability when it does so. Obviously, then, the dog's response to "Give me a paw" also represents equivalent language ability. The emitted language for both young infants and dogs is almost exclusively social in nature, attempting to elicit responses from other individuals. In dogs, the emitted language is actually a bit more complex than that of infants, since it emphasizes dominance and status relationships as well as the emotional states and desires of the communicator. Although a two-year-old may well try to manipulate others with displays such as temper tantrums, human children will not commonly attempt to communicate or express real social dominance until they are a few years older.

Some people have argued that because Doggish is mostly concerned with social and emotional issues, it really can't be classed as a true language. Such people, however, do not seem to really understand how humans actually use their language. Most of the time when we are talking, we are exchanging personal and social information. We are not usually having discussions about the philosophy of Aristotle, the theories of Einstein, or musings on the current state of the universe. We seem to be much more involved in the everyday aspects of our social lives.

Two British psychologists have actually sampled conversations to see what people are normally speaking about. Robin Dunbar took samples from all over England, while Nicholas Emler sampled ordinary conversations in Scotland.[1] Both found that more than two-thirds of our conversations are taken up with social and emotional matters. Typical topics dealt with who is doing what and with whom, and perhaps commentaries on whether that is a good or a bad thing. Other topics include who is moving up in the world and who is moving down and why. Many of the most emotional conversations had to do with how to deal with difficult social situations, and described complex interactions with lovers, children, fellow workers, neighbors, extended family, and so forth. Obviously, there were some complex technical discussions, perhaps triggered by a problem at work or a recently read book. Yet when I monitored over one hundred discussions between my colleagues at the university, I never found a technical discussion that went on for more than seven minutes without lapsing, at least for a while, into social conversation. In fact, only about a quarter of the time was spent on technical matters overall.

When we look at printed language, we get a similar picture. The best-selling books in the world are fiction. Most (even adventure or mystery stories) involve the description of characters in terms of their social interactions, their involvements with family, their personal ambitions, deceits that they have used and suffered, and, of course, sexual matters. The so-called romance novels continue to top the lists in volume of book sales. The only non-fiction category which claims a significant portion of the market is biography (and autobiography). It seems as though every actor, politician, sportsperson, broadcaster, and writer has written his or her story, and that there is an avid public that wants to read these accounts. But why do we buy these books? We don't read about the life of a politician to learn how to draft and pass legislation. We don't read about the life of a baseball player in order to learn how to hit the ball better, and we don't read about the life of an actor to learn how to memorize scripts. The reason that we read these books is because we want to know the social details. Who did they like or hate, how did they react to difficult social and emotional situations, whom did they interact with on their way to fame, and so on. The same pattern even carries over to newspapers. Roughly two-thirds of the actual printed "column inches" deal with human interest stories or social details designed to depict the intimate lives of the various celebrities and newsmakers being described. Much more time is spent on who is having conflicts or intimacies with whom, or who is "in" and "out" in the current social scene, than is spent on descriptions of objects or physical conditions in the world.

The fact that people spend most of their linguistic efforts in dealing with social and emotional topics does not cast into doubt their possession of language. Since Doggish seems to fit most of the other requirements of language, we should not deny that dog communication is language simply because it doesn't often deal with weightier topics than social interactions and emotional states. When my children were teenagers, I credited them with having language although it seemed to me that nearly all of their conversation dealt with their feelings and relations to others. In structure and complexity, the language of dogs is about equivalent to the language of a two-year-old child. The content of that language, however, is much like the content of two-thirds of adult human language, and is concerned with everyday social matters, the structure of the society, and the emotional world in which they live.

20

Talking Doggish and Doggerel

Up to now, most of our discussion has had to do with how we can understand what a dog is saying to us using canine language. Except for our brief look at receptive language in dogs, we haven't considered how people might talk to their dogs in a way that they will understand.

Most of us are already "talking" to our dogs in our own native tongue. This is not just the kind of talking that we do when we issue commands to tell the dogs to "Sit" or "Come." What I mean is when we speak to the dog in the same way that we might speak to another person or child. One survey suggested that 96 percent of all people talk to their dogs in this way. Nearly everybody admitted that they usually greet their dogs when they come home and also usually bid them farewell when they leave. Another common form of "conversation" involves complimenting the dogs by telling them that they are pretty or clever. Many people said they often tell the dog what they think of its behavior; this involves explaining that some recent behavior was stupid, naughty, helpful or funny. Sometimes they will extend the comment into a short narrative: "It's a good thing I found this mess before your mother did. She'd be very unhappy with you." Most people also admitted that they often ask questions of the dog about matters of interest to their canine companion, such as, "Do you want to go for a walk?" or, "Do you want a snack?"

One of the interesting aspects of human-canine communication is that the majority of dog owners claim they occasionally ask questions that the dog really couldn't be expected to answer (or even care much about), such as, "Do you think it's going to rain today?" or, "Do you think Sally will forgive me for what I said?" This conversation usually takes the form of a monologue, where the human being does all the talking while the dog provides just a friendly presence.

A somewhat more complex form of conversation takes the form of a dialogue where there is some give-and-take but only one speaker. In this kind of conversation, we usually look at the dog now and then, pausing at places where the dog might be expected to make a comment, and then continue on, as if the animal's silence had conveyed some meaning. If you listen to this kind of conversation, you hear something that sounds very similar to one side of a telephone conversation. A snippet might go: "What do you think I should give Aunt Sylvia for her birthday?" [Pause for a few seconds] "No. I gave her flowers last year. How about candy?" [Another brief pause] "Yes, of course it would be chocolates." [Pause] "I know what you mean. Dark chocolates with nuts in a pretty festive box. That's a really good suggestion, Lassie."

Another type of human and dog interaction is familiar to many dog owners but may seem a bit strange to an outsider. In this situation, the person not only talks to the dog but also gives the answers, essentially speaking the words that we think the dog would say in response to our comments. Thus, the person might say, "Well, Lassie, would you like a treat?" and when the dog responds by coming over, they might continue (often in another voice), "Of course I would, silly person!" You sometimes hear this kind of "conversation" when parents talk to young babies. In the expanded version of this, where the person provides both his or her own dialogue and that of the dog, it tends to produce a conversation much like the clichéd Hollywood movie sequence in which the schizophrenic carries on an argument among his various multiple personalities, each with a distinctive voice and its own character.

All of these forms of "conversation" are not really meant to communicate with the dog. Instead, their real function is to provide some social interactions for the human speaker, which might help to work through a problem, complete a thought, or explore an emotion. There is a lot of evidence to suggest that such interactions are important for psychological health. We generally get this social interaction from other people. However, elderly people, those who live alone, or any of us who find ourselves in an empty home when family and friends are away, can get some of the same benefits from talking to a dog. Some researchers have used blood pressure readings to show that it is actually less stressful talking about difficult matters with a dog than talking about them with a spouse. Other evidence shows that older people living by themselves are much less likely to fall into a depression or need psychological help if they have a dog as a companion and have the opportunity to engage in conversation with them.

Perhaps the strangest way that people talk to their dogs was described to me at a scientific meeting in Dallas. I was talking with a psychologist from Argentina who told me about how some people talk *to* their dogs and also to each other *through* their dogs. This was his description.

"There is a tribe in South America called the Achuar, who use dogs as a vital part of their communication. Among the Achuar, it is exclusively the women who care for the dogs. The dogs in turn guard the home. Some dogs help with household chores by carrying items in little basketlike packs that can be strapped to their backs. The women give the dogs their names and talk to them, much the way they talk to their children. However, the main task of dogs is to assist in hunting. Since hunting is exclusively the province of men, the dogs will spend long hours, sometimes even days, in the company of men. The men call the dogs by the names the women have given them. Men also train the dogs to hunt and to respond to the commands they will need to understand as part of the hunt. Sometimes, the men engage in idle conversation with their dogs, especially when they find themselves on a long, lonely trek after game, in much the same way that the women do when the dogs are around the home.

"I think that it's because both sexes share private time and conversation with dogs that these animals have come to be seen as occupying a place where the world of women and the world of men touch each other, but don't actually join. Because of this, the dog is available to play an important role in the Achuar system for reducing stress and strife between a man and his wife. Whenever there is some sort of potential conflict, a favorite dog will be brought into the house and asked to serve as go-between.

"It works like this. Suppose that I'm an Achuar man and my favorite dog is named Chuka. I might bring the dog inside and sit down and wait until my wife entered our home. Since I don't want to anger her by suggesting that she is a poor housekeeper and is not fulfilling her household duties, I might look directly at the dog and say something like, 'Chuka, perhaps you could speak to my wife, who loves you. There is to be a great feast in about a month and my dance cape has become old and worn. It's hard to dance well if you feel that others are looking at you and thinking you must be poor because your festival clothes look shabby.'

"My wife would not look directly at me, but might then turn to the dog and say, 'Chuka, because you know I love you, perhaps you could ask my husband if we have enough money to go to market this week and buy some shiny buttons or feathers. Tell him that if I had such things, I could make him a new collar for his dance cape so that he could look proud and rich at the feast next month.'

"Since both speak to, or through, the dog, they do not have to face each other. This means that they don't have to deal with facial expressions that might signal anger or hurt, which in turn might 'charge' the atmosphere between the couple. The dog doesn't seem to mind, and the messages seem to be carried quite accurately."

In most cultures, when we actually want to communicate with a dog, there seems to be a special form of language that we employ. We all know that our language changes under different circumstances. There is a formal language we use when we are talking to authorities or to an audience, that is more reserved and ceremonial than the language we use when talking with family and friends. Similarly, when we are writing, our sentences tend to contain more information and to use much more complex grammar and vocabulary than our spoken language. This explains why, when you read a written piece aloud, it often sounds artificial, convoluted, and pompous, not at all like conversational language.

Psychologists have found that there is also a special kind of language we use in talking to young children. This is a simplified language, often spoken in a singsong rhythm, with lots of repetitions. Sometimes we even use a higher-pitched tone of voice. Researchers have called this special language for children *Motherese,* since it is usually the language that mothers use when they are talking to their offspring. However, Motherese is certainly not confined to mothers, since virtually all adults, whether male or female, parents or not, tend to use it when talking to a very young child. Psychologists Kathy Hirsh-Pasek and Rebecca Treiman were able to show that the language we use when we are talking to dogs is very similar to Motherese.[1] They call this form of language *Doggerel.*

Doggerel is not the normal language that we would use around other adults. When we are talking to our canine companions, our sentences are much shorter. We will use utterances that have an average length of between ten and eleven words when speaking to an adult human, while when we talk to dogs the average statement length is reduced to around four words. We speak a lot more in imperatives or commands to our dogs, such as "Lassie, lie down," or, "Get off the sofa." Strangely, we also ask twice as many questions of our dogs than we do of humans, even though, as we saw earlier, we really don't seem to expect any answers. These questions are mostly trivial social exchanges rather than information seeking, for instance "How do you feel today, Lassie?" A lot of these queries are in the form of tag questions, which is where one makes an observation and then turns it into a question at the very end. An example would be, "You're hungry, aren't you?"

Doggerel is spoken mostly in the present tense, which means that we talk to our dogs about what is going on now rather than in the past or future. In fact, recordings show that around 90 percent of Doggerel is in the present tense, which is approximately half again as much as our normal speech to adult humans. Also, we are twenty times more likely to repeat ourselves than when we talk with people. These repetitions can be exact copies, partial repetitions, or some form of rephrasing. An example of rephrasing and repeating would be "Lassie, you're a good dog. What a good dog you are!" All of these characteristics of Doggerel are similar to Motherese.

There is one area where the way that we talk to our dogs and the way that we talk to our children is quite different, however. The difference between Doggerel and Motherese is clearest when it comes to *deixis,* which is the technical name for sentences that point out specific bits of information, such as "This is a ball" or "That cup is red." Sentences of this sort are usually viewed as an attempt to educate other individuals. Motherese contains a lot more sentences of this sort than does normal speech to adults, because mothers are actively trying to teach their children about language and the environment while speaking with them. Doggerel, on the other hand, contains only half as many such statements. Obviously, most of our speech to dogs is solely serving a social function for us; whether the dogs learn anything from it doesn't seem to be much of a concern.

One clear difference between Doggerel and any other speech we use is the tendency we have to occasionally mimic the sounds our dogs make. One evening I was visiting a friend when her poodle came up, stood in front of her, and gave a single huffing bark which sounded like "Woof." She replied, "Woof yourself, young lady. I'll feed you after company leaves." Her "Woof" in this response was a fair imitation of the dog's bark, and was clearly meant to be so. Mothers seldom imitate their child's random speech sounds, and to imitate the speech sounds or tones that another adult uses would probably be considered mocking or insulting. For some reason, imitating canine sounds is just another device we use to keep our conversations with our dogs going.

When we speak Doggerel, it sounds very different from the speech we use when talking to mature people. In addition to using a higher tone of voice, we also strongly emphasize intonations and any emotional phrasings. Furthermore, we may use a lot of diminutives—"walkie" for walk, "bathie" for bath. Words and phrases may be modified to make them appear less formal, as when we use words like "wanna" or "gonna." So, if you hear a woman ask in a singsong tone of voice, "Do ya wanna have a snackie?" you can probably safely infer that she is talking to her dog, although there is some slight chance that she is talking to a very young child. You can be reasonably certain she is not talking to a visiting adult friend.

Although there is no evidence that speaking to a dog in Doggerel helps the dog understand what we are saying, there is a lot of evidence which suggests that talking to dogs in a normal, purposeful, and meaningful manner improves their receptive language abilities. I am not referring to the casual conversations we have when we are simply "socializing" with our dogs. The educational form of conversation involves deliberately speaking to the dog, using simple phrases that anticipate activities of relevance to its life, such as "Let's go for a walk," or the question form, "Do you want to go for a walk?" When we are about to go up or downstairs with the dog, we say, "upstairs" or "downstairs." When we want the dog to follow us into another room, we say, "Let's go to the living room," and so on.

Since the purpose of this speech is to increase your dog's receptive vocabulary by increasing the number of words and signals that the dog knows, you should be consistent and always use the same words and phrases. For example, before giving the dog its food, you might announce "suppertime," "dinner is ready," "chowtime," "mess call," or, "Luncheon will be served in the main dining room." Which phrase you use doesn't matter as long as you select one word or phrase and use it consistently. Once the dog has the basic concept, it may be possible to introduce synonyms, but consistency builds up the dog's vocabulary faster. The idea is to get the dog to understand that specific human sounds predict specific events. Obviously, then, it is most productive if all the members of the family use the same words when speaking to a pet.

The dog will begin to show it is picking up new words in its receptive vocabulary by responding in an appropriate manner. After "Do you want to go for a walk?" the dog may move toward the door in anticipation; "Let's get your Frisbee" may cause the dog to dash to its toy box to find that throw toy. Each phrase begins to elicit an action in the dog which demonstrates that the word has been learned.

There are some simple tricks that may help to improve the dog's ability to understand language more quickly. Whenever you talk to the dog, remember to use its name. The dog's name comes to signal that the next sound has a meaning which applies to the dog. It is also important that each word has only one meaning. For example, if you use the word "out" when you want the dog to go out the door, then you should not use the same word when you want to remove an object from the dog's mouth. Perhaps the most useful teaching technique (especially with young dogs) is what I call *autotraining*. This helps the dog learn some basic commands without much effort.

Let us suppose that we are dealing with a puppy with the traditional name "Lassie." To autotrain words, you have to start by watching the dog's activities carefully as you interact with her. If you saw her begin to move toward you, you would say, "Lassie come," or when she begins to sit, say, "Lassie sit," After each of these actions, you would praise the dog, just as if it had responded correctly when you used the word as a command. What you are doing is attaching a label to an activity that the dog is already engaged in. Psychologists refer to this as *contiguity learning*. For many dogs, it takes only a few repetitions for the word to signify that action in the dog's mind. With this groundwork, it requires only a little additional effort to have the dog respond reliably to the word as a command.

Autotraining can make the learning of words involving simple actions much easier; it is particularly useful when you want to teach the dog the meaning of words that describe activities that are difficult to control directly. I use this technique when I am housebreaking my dogs. Each day, I walk them down a familiar route. As soon as the dog begins to squat to eliminate, I say, "Lassie, be quick," and repeat it once or twice during the elimination process. Afterwards, the dog is praised as if she had done something wonderful. Within a week or two, the phrase "Be quick" begins to have a meaning: when the dog hears it, she will begin to sniff around to choose a place to eliminate. A similar version of autotraining can teach a dog the meaning of the word "settle," which indicates that it is supposed to remain quiet, with little activity, in a particular region of the room or house. To teach a dog this word, wait until he's quiet and then say, "Rover, settle." Then walk over and quietly stroke him while you repeat the word "settle." It should not take many repetitions before the dog starts to show evidence that he understands the meaning. Soon, when he hears the word "settle," the dog will start to look for a comfortable place where he can sit or lie down, but still can watch the activities going on in the room.

If you have a dog that is sometimes shy around strangers, you can use autotraining to make it feel more secure. To do this, you need the help of a few friends and lots of treats. Walk the dog up to someone the dog doesn't know, then give the person a treat to give to her. Just before they offer the dog the treat, say, "Lassie, say hello," then repeat this as the dog takes the treat from the person's hand. After a few repetitions, the phrase "say hello" should come to mean that the person the dog is about to meet has a treat; this will produce a positive emotional response in the dog. Over time, the phrase develops a more general meaning for the dog: that the person they are about to meet is a friendly and non-threatening person (even if they don't happen to have a treat at the moment).

Everything we have said so far has been designed to help dogs to understand human language. However, if we are to have useful and meaningful communications with our dogs, we have to learn how to speak words in Doggish. We need to know also how to avoid inadvertently sending Doggish messages that might undermine our relationship with a dog.

An example of how important it is to use the right signals was provided by the French psychiatrist Boris Cyrulnik.[2] He studied communication between children and animals by carefully analyzing films and videotapes of their interactions. One of the things that most surprised him was that both groups of animals that he studied (dogs and deer) reacted more negatively and fearfully when interacting with normal children than they did when interacting with children with severe psychological problems, such as Down's syndrome or autism. He finally concluded that the problem lay in the signals these two groups of children were sending to the animals. When approaching a dog, Cyrulnik saw that the normal children looked straight at the animals. We have already discussed how staring is a threat in Doggish, so looking directly into the animal's eyes starts the encounter with a hostile message. Next, the children smiled at the dogs. The problem here is that the smile they gave was not a discrete upturning of the lips, but rather, a big, broad, open-mouthed smile. To the animal, this means that the children have just bared their teeth, clearly a mouth signal that suggests the threat of aggression. Typically, the children also raised their arms high and forward toward the dogs. In Doggish, this is the equivalent of a dog rearing up to make itself appear larger and more dominant, and threatening attack.

In most cases, the children also held out their hands with their fingers extended as they went to reach for the dog. Try this little experiment. Take one hand and extend the fingers forward, then turn that hand to the side and look at it. It looks much like an open mouth with teeth. Now turn the hand toward your face and look at it from the front. From the dog's point of view, this might not only be an open mouth, but one with very impressive, long teeth, and it is pointed directly at the dog. This is as definite a threat as you can give in Doggish.

Finally, after giving all of these threatening signals, normal children then usually rush directly toward the dog in a grand display of affection and enthusiasm. Unfortunately, for many dogs this is the last straw, the sign that an attack is beginning. Given these observations of children's behaviors, it should not be surprising that many children are bitten each year by dogs whose families describe them as normally friendly and non-aggressive. Perhaps the real surprise is that many more children are not bitten, given the hostile Doggish messages these youngsters are sending.

Cyrulnik found that children who had reduced mental capacities acted quite differently. They avoided looking directly at the animals, so there was no initial threat. They moved more slowly and often approached at an angle to the side, instead of face-on. Sometimes, they even moved in a sort of slow, sideways shuffle. When reaching out to touch the dogs, they tended to keep their arms low, and their fingers usually curled inward. Thus, the very nature of their disorders cause them to be non-threatening to the animals.

In one case that Cyrulnik observed, a pair of dogs were eating from a dish when they were approached by both a normal and a mentally handicapped girl. The normal girl came first, reaching toward the dogs; she received a threatening growl, which caused her to back off immediately. The mentally deficient child, however, never looked directly at either dog. She crawled along and pushed the dogs away by butting their rears with her head, in much the same way that puppies do to block the aggressiveness of adult dogs. This allowed her to come up quite close, at which point she lay down, and then gently stole the dish away. The dogs put up with this behavior because there were no threat signals and no signals associated with assertions of adult social dominance.

These examples show us that dogs do read human body language as though it were Doggish, and this suggests that humans can deliberately use these Doggish signals to communicate with canines. Suppose you encounter a shy or nervous dog that you want to make friends with. If you see the dog acting in a fearful manner, you should immediately turn your head and eyes away and look in another direction. Then slowly angle your body so that your side is facing toward the dog. All of your movements should be slow and casual. Don't walk directly toward the dog, but rather in a slanted direction, as if you were going to pass right by him, and remember to always keep your side toward the dog. When you are close, but not so close that you are raising the dog's anxiety level, you should kneel down. It sometimes helps to look as if you are interested in something on the ground, perhaps touching the floor in front of you. You might look at the horizon, or off to the side, but never look directly at the dog. Now slowly take out a treat, place it in your cupped hand, and hold your hand a short distance out to the side. Your relationship to the dog should be something like that shown in the top section of Figure 20-1.

At this point, I like to talk in a quiet manner, making soothing sounds

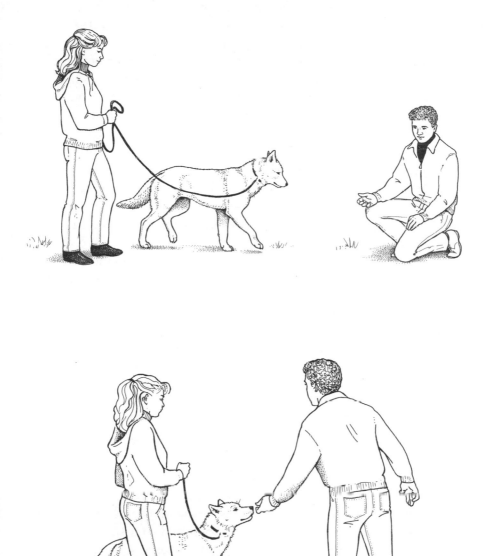

Figure 20-1 The top figure illustrates how to greet a dog who is showing shyness or anxiety, while the bottom figure demonstrates how to greet an unfamiliar dog who is showing no fear or defensiveness.

in a slightly higher-pitched voice than I usually use. I also try to use the dog's name if I know it. This seems to have a calming effect. Usually after only a few moments, the dog will approach you. Even if you feel a cold nose against your hand, don't turn your head yet. Wait until the dog has taken the treat and then slowly turn your head a bit. A second treat can be offered as you look in the general direction of your hand. Don't rush things, and don't try to pet the dog until it accepts your closeness. It should only take a minute or so to work through these steps.

Even if a dog is not showing any fear at your approach, when the dog is unfamiliar to you, it is always best to begin your greeting ceremony by turning your side to the dog. You should avoid looking directly at its face but glance off toward the horizon. If you offer a treat or a sniff, do this by holding your hand out at a distance from your side, and make sure that your fingers are curled in and together. This typical greeting position is shown as the bottom section of Figure 20-1. As in the case of the obviously shy dog, a bit of quiet talking and use of the dog's name always seems to help.

Even the simple act of petting the dog has Doggish meanings. If you move your hand toward the dog so that it is higher than the dog's head, this could be interpreted as a potential dominance signal, similar to rearing up or placing a paw over another dog. When petting a dog, your hand should start low, stroking the chest first, then working your way toward the head in order to avoid any dominant or challenging signal.

Suppose that you are actually threatened by a dog. If the dog is showing the full threat pattern—mouth open, teeth bared, gums exposed, hackles raised—you will have to find some way to communicate to it that you are no threat. It doesn't matter whether the dog is acting in an aggressive manner because it is dominant and confident and feels challenged for some reason, or whether it is threatening because it is frightened and insecure. Even if the tail position and ears suggest that the aggression is motivated by fear, you should not relax, since more people are bitten by insecure and frightened dogs than by dominant animals.

When a dog is signaling a threat, the first thing to remember is that you should *not* turn and run, since this will trigger the dog's pursuit response. Your reaction to this signal should be to cast your gaze slightly down and to the side, and blink once or twice. These are submissive pacifying responses. Open your mouth a little, which indicates a bit of a counterthreat, suggesting that you will respond to any aggression if the dog initiates it. Next, take a few slow steps backward, never glancing directly into the dog's eyes. If you can manage to control your breathing enough, turn your head a bit further to the side and see if you can force yourself to yawn, or to say something soothing in a high-pitched tone of voice. When you are a good distance away, turn so that your side is toward the dog. If he moves toward you, face him again, then give another exaggerated set of eye blinks, look to the side and down, and continue to back off slowly. If he shows no increase in his arousal and no threatening change in behavior when you turn your side toward him, you can *slowly* walk away. Keep your eyes from directly meeting the dog's and try to move as smoothly as you can.

Some people have suggested that the principal way to prevent aggression and to ensure obedience in your dog is to use aspects of dog behavior to make sure that your dog knows that you are the "leader of the pack." The idea was to use such dominant and punitive measures that the dog would never challenge your authority. In earlier times, perhaps until the 1920s, dog obedience training was known as "dog breaking." Between the nineteen-thirties and nineteen-fifties, it was still possible to get "dog whips" and leashes shaped so that the handle could be used as a whip. In response to increasing public concern about cruelty to animals, the dog whip was replaced by the choke chain and the correction jerk on the leash.

As we began to learn more about the behaviors of wild and domestic canines, some dog trainers began to suggest that dogs could be punished for any challenge to human leadership by using the same behaviors that more dominant canines do in the wild. Unfortunately, their attempts to use Doggish signals were often very inaccurate. For example, some people looked at the way that adult dogs responded to each other when they got angry. If the conflict could not be resolved, one member of the feuding pair might bite the other on the nose or on the ears. It was suggested that to enforce dominance, the human master should bite dog in the same way. Attempting to bite a medium- or large-sized dog on the nose is simply madness. If the dog is angry, its mouth and muzzle are much better designed to do damage than those of a human. I, for one, would never offer my face as an easy target by attempting to bite a large, angry dog on the nose! Biting the dog on the ear is equally misguided. Again, the dog can easily turn its face and respond with its larger, sharper teeth. In addition, it is possible to permanently damage or disfigure the ear. Also, there are legal consequences, since some people have been prosecuted under animal cruelty statutes when caught biting their dogs. The worst part of this strategy is that it just doesn't work. Biting is what occurs when communication breaks down. It is the last resort, when Doggish signals have failed to resolve a conflict. It is not, in itself, a communication signal.

The well-known ethologist Konrad Lorenz suggested that we can discipline puppies by grabbing them by the scruff of the neck and shaking them. This was based on the observation that sometimes a mother dog would do this with an unruly puppy. More modern trainers have taken this step further and suggested that we do this to adult dogs to inform them that challenges to our dominance will not be accepted. If the dog is large enough, one is instructed to grab the loose skin on either side of the neck, stare into its face, and shake it violently. This maneuver will terminate violence, but not because it is a Doggish signal. Rather, it constitutes a greater level of violence and effectively "wins the fight." This is enforcement, not communication.

More recently, some dog trainers have suggested using the "alpha roll." They observed, quite correctly, that a submissive dog will often signal its lower status and its willingness to submit to a more dominant dog by rolling on its back and exposing its underside. Therefore, they reasoned, we can use this Doggish signal to assert that we are pack leaders and dominant. The idea was to forcibly roll the dog on its back, and if it tried to move, to hold it down and growl at it. While the interpretation of the Doggish signal is correct, the strategy is not. When dogs interact, you never see the dominant dog actually force the submissive dog onto its back. The submissive dog does this on his own, *after* he has acknowledged the dominance of the other dog. Forcing the dog onto its back is the equivalent of an abusive parent beating a child to force it to say, "I love you." Although he or she may have forced the words out of the child's mouth, they cannot force the statement to be true. The child may say the words but still hate the parent. Forcing a dog into a submissive position is the Doggish equivalent of this scenario. Even worse, this technique may actually anger the dog enough to provoke it to attack.

Forcing a dog into an alpha roll, or shaking the dog, both constitute actual physical aggression. Physical aggression is not communication. If there is good communication, then such confrontations need not occur.

Having your dog under full control is achieved by a combination of two factors. The dog must accept that you are the Alpha dog and the animal must *want* to please you. This requires a balancing act in your messages. You must communicate that you are the pack leader and dominant, but you also must assure the dog that it is accepted and has the right to enjoy an peaceful life as a member of "your pack." Although this is not the place to discuss the full issue of canine dominance, there are some simple rules which assure that the dog understands who the pack leader is. The Alpha dog controls the resources, whether these are food or the opportunity to play or whatever. You should never give the

dog anything "for free." Simply requiring the dog to do something before it gets what it wants, even if this is as simple as requiring it to sit or
lie down before it gets a treat or pat on the head, communicates dominance without signaling threat or aggression. What the dog is learning in
these situations is that it must respond to your messages. As the leader,
you will reward him with the things he wants. If you feel that you must
"shout" the message that you are leader in Doggish, then simply have
the dog stand or sit next to you and rest your hand or your arm over the
dog's shoulders. This is the equivalent of a dog asserting dominance by
placing its head or paw over the shoulders of another. If your dog resists
this signal, then it really doesn't believe that you are the leader yet.

Up to now, we have concentrated completely on speaking to the dog
and getting the dog to speak to us in return. What about when we want
the dog to *stop* speaking? For example, I was watching a beginners' dog
obedience class when a border collie named Richard began barking at
the other dogs sitting in line across the room. Normally, I don't have
much concern about the occasional bark from a dog. Nonetheless, his
excited barks were getting very loud, and in that small room, they were
becoming annoying. Richard's owner was now frantically yelling, "No!
Stop that!" Unfortunately, that was a bad move.

Here we have a situation where the dog's master simply doesn't understand the basics of Doggish. To a dog, loud short words like "No!",
"Shut up!", "Don't bark!" sound just like barks. Think of it this way.
The dog barks to signal a potential problem. Now you (who are supposed to be leader of the pack) come over and bark too. This clearly indicates that you agree that this is the right time to sound the alarm.
Richard read the situation this way and was now barking in an almost
frenzied manner.

With the noise level escalating, other members of the class were looking around hoping something could be done to stop the commotion. The
instructor (call him George) responded to this pressure. He had clearly
learned a little bit about dog communication, and decided to use a dominant threat to stop the clamor. His attempt to quiet the dog involved
staring directly into his eyes in an accusing manner. Richard's ears folded
back submissively and he lowered his body to show that he recognized
the threat. His barking stopped. Unfortunately, this moment of quiet did
not last very long. Actually, it only lasted until George looked away.
Once eye contact was broken, Richard started barking again. Now
George looked aggravated. Instead of dealing with barking as a communication, he decided to look at this as a "situation" that needed training
and immediate correction.

George's next attempt was to take the dog and place him in a sitting position next to his left leg. The moment that Richard barked, George's right hand shot up under Richard's muzzle, applying a sharp smack which clapped the dog's jaws together for a moment, and then, just as quickly, the hand returned to George's side. The scene was repeated a couple of times—bark, smack, silence—bark, smack, silence. When Richard was quiet again, George returned to the front of the class. Of course, the moment George was out of smacking range, Richard began to bark again.

In order to stop dogs from barking, many different techniques have been tried. I have seen water pistols and squirt bottles, lemon juice sprays, muzzles, adhesive tape, rolled magazines, rattle cans, and electric collars used. Sometimes they work—often not. Even when they do work, they tend to be harsh, and can damage the relationship between dog and master. The dog is barking to communicate something that it feels is relevant to the pack. He might be sensing danger and attempting to warn his packmates. He might be sensing an incursion into the pack's territory and feel that he is defending the homestead. Whatever the reason, he feels that he is responding for the good of his loved ones. Imagine what goes through the dog's mind when this act of devotion is met by violence. It would be much like what a person might feel if they spotted smoke in a building, went to warn friends that they should evacuate, only to be punched in the face and told to shut up. Such aggressive actions are bound to damage future relationships. Furthermore, these aggressive "corrections" provide only a relatively short-term solution to a problem that is easily solved if you understand dog communication patterns.

We already know that although wild canines do not bark much, they do bark as puppies. In the safety of the den area, there is little harm in such noise; but as the puppies grow older and begin to accompany the adults on hunts, such barking becomes counterproductive. A wolf puppy or adolescent who barked at an inappropriate time could easily alert potential prey that the pack is near. The barking could also attract the attention of other, larger predators, who might have developed a taste for wolf meat. To stop this, a simple communication pattern has evolved. It obviously does not involve any loud sound signal, since the major aim of the behavior is to stop the noise. This means that a wolf will not stop another's barking by barking back at it. The signal to stop should also not involve direct aggression against the noisy individual. Nipping or biting the barker is apt to cause yelps of pain, growls, or dashing about to avoid or respond to the physical violence. All of this noise and thrashing around would be just as likely to alert other animals as the original barking itself. Therefore, the method demanding quiet should itself be relatively quiet and not physically aggressive.

The procedure worked out by wild canines to stop barking is really quite simple. The pack leader, the puppy's mother, or any pack member who is obviously higher in dominance ranking can give the signal for silence. To quiet barking, the dominant animal places its mouth over the offender's muzzle, without actually biting, and then gives a short, low, and breathy growl. The low growl will not be heard very far and it is short in duration. The mouth over the muzzle is not actually inflicting pain, so there is no yelping or attempt to escape. Silence usually follows immediately. This maneuver can be seen in Figure 20-2.

Humans can mimic this behavior as a simple way to tell a dog to stop barking when it is near you. With your dog sitting at your left side, slip

Figure 20-2   The signal that an adult dog uses to tell a puppy to stop barking.

the fingers of your left hand under the collar at the back of the dog's neck. Pull up on the collar with your left hand, while your right hand folds over the top of the dog's muzzle and presses down. In a quiet, businesslike, and unemotional tone, you simply say, "Quiet." You repeat this silencing maneuver whenever it is necessary. Depending upon the breed, it may take anywhere from two to a couple of dozen repetitions to associate the calmly stated command "Quiet" with an end to barking.

What you have done in this instance is to effectively copy the way in which the pack leader will silence a noisy puppy or other pack member. Your left hand on the collar simply immobilizes the head. Your right hand serves the same function and communicates the same message as the leader's mouth over the noisy animal's muzzle. The softly spoken "Quiet" mimics the short, low, breathy growl.

Returning to the obedience class and the barking border collie, I signaled to George that I would silence the dog's din. Richard was in full frantic barking mode when I arrived beside him. I used the hushing signal described above, and in a low voice said "Quiet." Richard required only three repetitions of this action to end his barking for the evening. I later learned from his handler that within a week, a low, matter-of-fact "Quiet" became all that was needed to stop his barking.

Be sure, however, that you use this procedure to stop a dog from barking only when barking is unnecessary, as in an obedience class or a public place. Remember that we specifically bred dogs to bark, so if your dog sounds the alarm at the approach of a stranger, or even at the sight of a cat outside your window, don't correct him. If there is no cause for any action, just call him to your side and give him a quick pet or a rub. By barking, your dog is only doing the job we designed him to do thousands of years ago.

It is probably best to leave the barking to the dogs, and for humans not to deliberately try to imitate these sounds. Linda Cawley, a lawyer, tells of once having to defend a case involving nuisance barking. The individual charged, however, was a human, not a dog. The situation arose in Lakewood, Colorado. The defendant was a man who was in his backyard when his neighbor's dog started barking at him. The man obviously did not understand the principle among dogs that "barking begets more barking." He thought he could silence the dog by barking back at it. The dog barked, he barked back, the dog barked more vigorously, the man barked louder, and what was described as a virtual barking war broke out. This kind of behavior continued over a number of days, and the dog's owners became upset. Rather than doing anything to quiet their dog, they brought charges against the man. It seems incredible, but this man was actually charged with animal cruelty for harassing the dog! Nobody wanted to back off in this fight. The dog continued to bark, the man continued to bark, and the wheels of justice turned. The case actually went to court. Cawley, who specializes in animal law, was called in to defend the man.

"I fought the charge as a freedom of speech issue," Cawley said. The man had the right to express his opinions, and she felt that the language he used to do so, in the privacy of his own backyard, was irrelevant. The judge agreed and gave a not guilty ruling, confirming people's right to bark as a means of personal expression.

Postscript

ONE LAST WORD

There is one sound that dogs make which I have not included in my discussion of Doggish vocalizations. I didn't include it because it is an automatic sound, which probably was not intended by either evolution or the gods to be communication at all, but it has come to mean something to me. It is the sound of dogs breathing.

At night, when I lie down to sleep, my old dog Wiz lies on the bed beside me, while Odin lies on a cedar chip pillow on the floor close by my head. Just across the room, my puppy, Dancer, who is not quite fully house-trained, sleeps in his wire kennel. In the quiet and the darkness, sounds are amplified. I can hear the low, slow breathing of the big black dog, the short breaths of the orange puppy, and the occasional sniffle and snore of the old white dog. As I listen to those soft sounds, I think of some earlier man, lying in a cave or rude shelter, resting on a bed of hides or straw. It was a hostile, dangerous world. Weapons were primitive, resources often sparse, and there were menacing things that moved in the night. That long-gone ancestor also had dogs who lay beside him as he tried to sleep. His dogs breathed these same sounds and these sounds had meaning. They were not merely part of the language of nature—they were the sounds of safety and comfort, a recitation of the dog's eternal contract with humans.

"I am here with you," the dog's breath said. "We will face this life together. There is no beast or intruder that can steal up on you undetected because I am here, and I will be your eyes and ears. No harm will come to you because I am at your side to warn you, and to defend you if need be.

"We will hunt together tomorrow. We will herd together tomorrow. We will share the sunshine tomorrow. We will explore this world together. We will laugh together. We will play together, even though neither of us is any longer a child.

"If luck turns bad, then when you grieve, I will comfort you. You will never need to be alone again. I promise this. As your dog, I will sing this promise to you, and whisper it to you at night, every night, with my breath."

I can hear these words in my dogs' soft sounds of breathing, and, just like my ancient ancestor, I understand these words and I am comforted. In my heart I know that if the language of dogs were so limited that this was the only message they could send, it would still be enough.

Appendix

VISUAL GLOSSARY AND DOGGISH PHRASEBOOK

In this book, I have tried to assemble a variety of different signs and signals you can use to interpret what your dog is saying. This Appendix gathers together the main themes and meanings of canine communication. There are two sections. First is a Visual Glossary, which shows pictures of some typical postures and facial expressions so that, when you see them, you can get an immediate overview of what the dog is communicating. Second is a Doggish Phrasebook, which takes each of the various major signals—based upon sounds, facial expressions, eye and ear signals, tail positions, and general body language—and interprets them in everyday human language terms. In addition, under the heading "Conditions and Emotions," I have tried to give you insight as to the underlying psychological state, or the circumstances and events that might trigger these signals. Together, these may help you to better understand what your dog is trying to say to other dogs and to people.

Visual Glossary

Relaxed

- Ears up (not forward)
- Head high
- Mouth open slightly, tongue exposed
- Loose stance, Weight flat on feet
- Tail down and relaxed

This array of signals communicates a relaxed, reasonably content dog who is unconcerned and unthreatened by any activities going on in its immediate environment.

Alert and Attentive

- Ears forward (may twitch as if trying to catch a sound)
- Eyes wide
- Smooth nose and forehead
- Tail horizontal (not stiff or bristled)
- Mouth closed
- Tail may move slightly from side to side
- Slight forward lean standing tall on toes

When something of interest is encountered or enters the environment, these signals communicate that attention is now being paid to them and the dog has entered a state of alertness.

Dominance / Aggression
(Offensive threat)

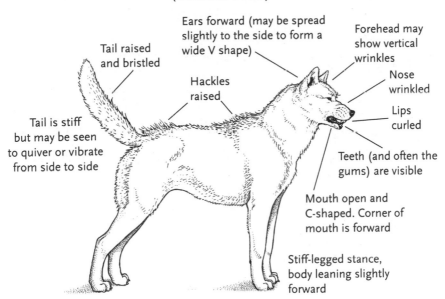

Ears forward (may be spread slightly to the side to form a wide V shape)

Forehead may show vertical wrinkles

Tail raised and bristled

Hackles raised

Nose wrinkled

Lips curled

Tail is stiff but may be seen to quiver or vibrate from side to side

Teeth (and often the gums) are visible

Mouth open and C-shaped. Corner of mouth is forward

Stiff-legged stance, body leaning slightly forward

These signals are given by a very dominant and confident animal, who is communicating both its social dominance and threatening aggression if it is challenged.

Fear / Aggression
(Defensive threat)

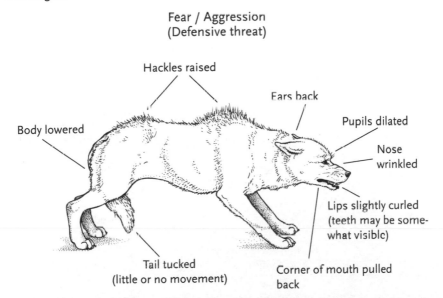

Hackles raised

Ears back

Pupils dilated

Body lowered

Nose wrinkled

Lips slightly curled (teeth may be somewhat visible)

Tail tucked (little or no movement)

Corner of mouth pulled back

This set of signals communicates that the dog is frightened but is not submissive and may attack if pressed. These signals are addressed directly toward the individual who is threatening.

Stress and Anxiety

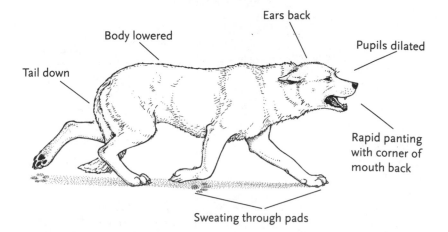

This is the pattern of signals which communicate that a dog is under stress. The source of the stress may be social or environmental, and the signals are not being specifically addressed to any other individual.

Fear / Submission
(Active submission)

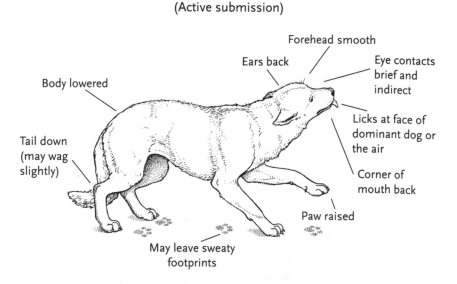

This pattern of signals communicates that the dog is somewhat fearful and is offering signs of submission. Most of these signals are designed to pacify the individual who is of higher social status in order to avoid any further challenges or threats.

Extreme fear / Total submission
(Passive submission)

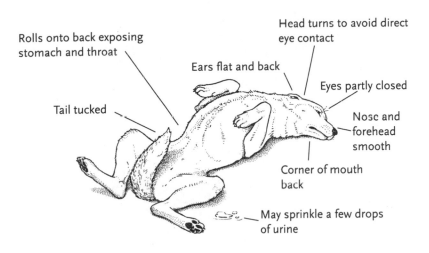

Rolls onto back exposing
stomach and throat

Head turns to avoid direct
eye contact

Ears flat and back

Eyes partly closed

Tail tucked

Nose and
forehead
smooth

Corner of mouth
back

May sprinkle a few drops
of urine

This pattern of signals indicates total surrender and submission. The dog indicates its lower status and grovels before the higher-ranking animal to pacify it and avoid confrontation.

Playfulness

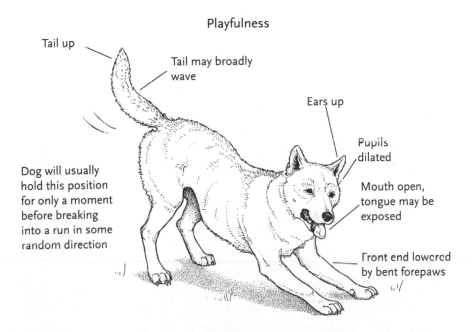

Tail up

Tail may broadly
wave

Ears up

Pupils
dilated

Dog will usually
hold this position
for only a moment
before breaking
into a run in some
random direction

Mouth open,
tongue may be
exposed

Front end lowered
by bent forepaws

This is the basic invitation to play. It may be accompanied by excited barking, or playful attacks and retreats, and may be used as a "punctuation mark" to indicate that any previous rough behavior was not meant as a threat.

A Doggish Phrasebook

This collection of the major signals used by dogs in their communication is not complete, since there are so many shadings of meaning. It is arranged by signaling system. Thus, there are separate sections for sound signals (broken into barks, growls, whines, and whimpers) and visual signals (broken into eye signals, yawns, facial signals, tail signals, and body language). In many cases, one signal must be read in conjunction with others to determine the meaning unambiguously.

I have given a common everyday human language equivalent for most phrases. These are provided to give the flavor of the communication, but in many instances, other related phrases could work as well. The section on "Conditions and Emotions" will help flesh out those alternative meanings and give you a sense of the feelings and events that trigger communication.

In addition, I have provided some "General Rules" within the sections. All of the signals can be modified to add aggressive content, or pacifying and calming content, or to indicate greater excitement and arousal behind the message.

All these phrases are from canine productive language and represent signals that the dog makes which we, as humans, can interpret. The aspects of human language that a dog may be able to understand depend upon individual experience with human speech and body language, and thus cannot be listed here.

I hope this helps you and your dog to reach a mutual understanding.

SOUND SIGNAL	HUMAN LANGUAGE MEANING	CONDITIONS AND/OR EMOTIONS
BARKS		
Rapid strings of 3 or 4 barks with pauses between (midrange pitch)	"Gather together. I suspect there may be something we should look into."	Alerting call suggesting more interest than alarm
Rapid barking (midrange pitch)	"Call the pack!" "Someone is entering our territory." "We may need to take some action soon."	Basic alarm bark. Dog is aroused but not anxious. Initiated by approach of a stranger or occurrence of unforeseen event. More insistent than the broken bark above
Continuous barking (a bit slower and lower pitch)	"An intruder [or danger] is very close." "Get ready to defend yourself!"	A more worried form of the alarm bark, which senses imminent threat
Long string of solitary barks with pauses between each	"I'm lonely and need companionship." "Is there anybody there?"	Usually triggered by social isolation or confinement
One or two sharp short barks (high or midrange pitch) spoken quietly	"Hello there!" "I see you."	Typical greeting or acknowledgment signal, initiated by arrival of, or sight of, a familiar person
Single, loud, sharp short bark (lower-midrange pitch)	"Stop that!" "Back off!"	Annoyance bark when disturbed from sleep, hair is pulled, etc.
Single moderately loud, sharp short bark (higher-pitched)	"What's this?" "Huh?"	Sign of being surprised or startled
Single bark, more deliberate in delivery, not as sharp or short as above (mid- to upper-midrange pitch) which may sound a bit forced or artificial	"Come here!"	Often a learned communication, which tries to signal a human response such as opening a door, giving food, etc.

SOUND SIGNAL	HUMAN LANGUAGE MEANING	CONDITIONS AND/OR EMOTIONS
BARKS (continued)		
Stutter-bark (e.g., "ar-Ruff!")	"Let's play."	Usually given with front legs flat on the ground and rear held high as play invitation
Rising bark	"This is fun!" "Let's go!"	Excitement bark during play or in anticipation of play, as in the master throwing a ball

General Rules for barks:
—Low-pitched means dominant or threatening, high-pitched indicates insecurity or fear.
—The faster the rate, the more aroused and excited.

SOUND SIGNAL	HUMAN LANGUAGE MEANING	CONDITIONS AND/OR EMOTIONS
GROWLS		
Soft, low-pitched (seems to come from the chest)	"Back off!" "Beware!"	From a dominant dog who is annoyed or is demanding that others should move away
Growl-bark (low-pitched "Grrrrr-Ruff")	"I'm upset, and if you push me, I'll fight!" "Packmates, rally round me for defense!"	A somewhat less dominant sign of annoyance with a suggestion that help from pack members would be appreciated
Growl-bark (higher midrange pitch)	"You frighten me, but I'll defend myself if I have to!"	A worried threat from a dog who is not confident but will use aggression if pressed
Undulating growl (pitch rises and falls)	"I'm terrified!" "If you come at me, I may fight, but I also may run."	The fearful-aggressive sound of a very unsure dog

General Rules for growls:
—Low-pitched means dominant or threatening, high-pitched indicates insecurity or fear.
—The more frequently the pitch and steadiness of the growl varies or changes, the more unsure the dog is.

SOUND SIGNAL	HUMAN LANGUAGE MEANING	CONDITIONS AND/OR EMOTIONS
HOWLS AND BAYING		
Yip-howl (e.g., "yip-yip-yip-howl," with the howl prolonged)	"I'm lonely." "Is there anybody there?"	Triggered by isolation from family and other dogs
Howl (often sonorous and prolonged)	"I'm here!" "This is my territory!" "I hear your howls."	Dogs use this to announce their presence, to socialize over a distance, and to declare territory. Although it may sound sad to a human, the dog is quite content
Bark-howl (e.g., "Ruff-ruff-howl)	"I'm worried and alone." "Why doesn't somebody come to be with me?"	The mournful sound of a dog that is lonely and isolated but fears nobody will respond to its call
Baying	"Follow me!" "All together now!" "I've got the scent, so keep close!"	A hunting call from a dog that has the scent, is tracking the quarry, and is ensuring that pack-mates are nearby for assistance
WHINES, WHIMPERS, MOANS, AND CRIES		
Whining that rises in pitch at the end of the sound (may sound like it is mixed with a bit of a yelp)	"I want . . ." "I need . . ."	A request or plea for something: louder and more frequent indicates strong emotion behind the plea
Whining that drops in pitch at the end of the sound or simply fades with no pitch change	"Come on now, let's go!"	Usually indicates excitement and anticipation, as when waiting for food to be served or a ball to be thrown
Moan-yodel (e.g., "Yowel-wowel-owel-wowel") or Howl-yawn (e.g., a breathy "Hooooooo-ah-hooooo")	"I'm excited, let's do it!" "This is great!"	Pleasure and excitement signals when something the dog likes is about to happen; each dog will settle on one of these sounds

SOUND SIGNAL	HUMAN LANGUAGE MEANING	CONDITIONS AND/OR EMOTIONS
WHINES, WHIMPERS, MOANS, AND CRIES (continued)		
Soft whimpering	"I hurt." "I'm really frightened."	A fearful passive/submissive sound that occurs in adults as well as puppies
Single yelp (may sound like a very short, high-pitched bark)	"Ouch!" (or some short profanity)	A response to a sudden, unexpected pain
Series of yelps	"I'm really scared!" "I'm hurting!" "I'm out of here!" "I surrender!"	An active response to fear and pain, usually given when the dog is running away from a fight or painful encounter
Screaming (may sound like a child in pain combined with a prolonged yelp)	"Help! Help!" "I think I'm dying!"	A sign of pain and panic from a dog who is fearful for its very life
Panting	"I'm ready!" "When do we start?" "This is incredible!" "This is tense!" "Is everything okay?"	Simple sound of stress, excitement, or tense anticipation, which may be accompanied by damp footprints on the floor
Sighs	"I'm content and am going to settle down here awhile." "I'll give up now and simply be depressed!"	Another simple emotional signal, which terminates an action. If the action has been rewarding, it signals contentment; otherwise it signals a termination of effort
EAR SIGNALS (usually read in conjunction with other signals)		
Ears erect or slightly forward	"What's that?"	Sign of attention
Ears definitely forward (combined with bared teeth and wrinkled nose)	"Consider your next action carefully—I'm ready to fight!"	The active, aggressive challenge of a dominant and confident dog

VISUAL SIGNAL	HUMAN LANGUAGE MEANING	CONDITIONS AND/OR EMOTIONS
EAR SIGNALS (continued)		
Ears pulled back flat against head (combined with bared teeth and wrinkled forehead)	"I'm frightened but I'll protect myself if you appear to be trying to hurt me."	A fearfully aggressive signal from a non-dominant dog who feels threatened
Ears pulled back against head (teeth not visible, forehead smooth, body held low)	"I accept you as my strong leader." "I know you won't hurt me because I'm no threat."	An active pacifying and submissive signal
Ears pulled back against head (tail held high, blinking eyes, and relaxed open mouth)	"Hello there. We can have fun together."	A friendly gesture, often followed by mutual sniffing or invitation to play
Ears pulled slightly back to give impression of a slightly splayed or sideward spread of the ears	"I'm suspicious about what's going on." "I don't like this and may fight or run."	A sign of tension or anxiety about the situation that is unfolding; this may quickly turn into either aggression or fear depending upon what happens next
Ears flickering, usually slightly forward, then a moment later slightly back or downward	"I'm just looking this situation over, so don't take offense."	A submissive and pacifying signal from a dog who is undecided and perhaps a bit apprehensive
EYE SIGNALS		
Direct eye-to-eye stare	"I challenge you!" "Stop that now!" "I'm boss around here, so back off!"	An active dominant/ aggressive signal, usually given by a confident dog who is having a social confrontation with another
Eyes turned away to avoid direct eye contact	"I don't want any trouble!" "I accept the fact that you are boss around here."	A signal of submission, with some undertones of fear

VISUAL SIGNAL	HUMAN LANGUAGE MEANING	CONDITIONS AND/OR EMOTIONS
EYE SIGNALS (continued)		
Blinking	"Okay, let's see if we can avoid a challenge." "I'm not really threatening you."	Blinking adds a pacifying gesture to the threat stare and lowers the level of confrontation without giving up much status

General Rules about eye signals:
—The larger the pupil size, the more emotional and aroused the dog is.
—The larger and rounder the eye shape, the more dominant and threatening the signal.
—The smaller the eye appears (and the closer to being closed), the more pacifying and submissive the signal.
—When movements are seen in region of the dog's forehead that corresponds to where our eyebrows would be, the emotional signals are virtually the same as for human eyebrow signals.

FACIAL SIGNALS (may need to be interpreted along with other signals)		
Mouth relaxed and slightly open (tongue may be visible or even slightly draped over lower teeth)	"I'm happy and relaxed."	The closest dogs come to a human smile
Mouth closed (no tongue or teeth visible, dog looks in a particular direction, leaning slightly forward)	"This is interesting." "I wonder what's going on over there?"	A sign of attention or interest
Lips curled to expose some teeth (mouth still mostly closed)	"Go away! You're bothering me!"	First sign of annoyance, menace, or threat; may be accompanied by a low, rumbling growl
Lips curled up to show major teeth; some wrinkling of the area above the nose; mouth partly open	"If you press me, or do anything I may interpret as a threat, I'll fight."	Active aggressive response, which may be motivated either by a challenge to social dominance or by fear
Lips curled up to expose, not only all the teeth but also the gums above the front teeth; visible wrinkles above nose	"Back off now—or else!"	High level of active aggression, with high likelihood that failure to give the dog additional space will result in an attack

VISUAL SIGNAL	HUMAN LANGUAGE MEANING	CONDITIONS AND/OR EMOTIONS
FACIAL SIGNALS (continued)		
General rules about face signals involving the mouth: *—The more exposed the teeth and gums are, the greater the signaled threat.* *—If the mouth is open wide and C-shaped, the threat is based on dominance.* *—If the mouth is open but the rear corner seems to be pulled back, the threat is based on fear.*		
Yawns	"I'm a bit tense right now."	Simple sign of stress or anxiety; may also be used to diffuse a threat
Licking the face of a person or dog	"I'm your servant and friend and recognize your authority." "I'm hungry. Do you have a snack?"	A pacifying gesture of active submissiveness, acknowledging the dominance of another. As a holdover from puppyhood, this is also a food request
Licking the air	"I bow before your authority and hope you will not hurt me."	An extreme pacifying gestures, showing fearful submissiveness
TAIL SIGNALS		
Tail horizontal, pointing away from the dog but not stiff	"Something interesting may be happening here."	Sign of relaxed attention
Tail horizontally straight out, pointing away from the dog	"Let's establish who's boss around here."	Cautious greeting ritual and initial mild challenge to a stranger
Tail up and slightly curved over the back	"I'm top dog around here and everybody knows it."	Confident signal from a dominant dog
Tail held lower than the horizontal but still some distance off the legs, perhaps with an occasional relaxed swishing back and forth	"All is well." "I'm relaxed."	Normal stance of an unconcerned dog with no particular worries at the moment

VISUAL SIGNAL	HUMAN LANGUAGE MEANING	CONDITIONS AND/OR EMOTIONS
TAIL SIGNALS (continued)		
Tail down, near hind legs, may wag slowly over a short distance; legs straight and body at normal height	"I'm not feeling well." "I'm a bit depressed."	Sign of physical or mental distress or discomfort
Tail down, near hind legs, with a lowered body position caused by bending of the legs	"I'm feeling a bit insecure."	Sign of social anxiety and mild submission
Tail tucked between the legs	"I'm frightened." "Don't hurt me!"	A submissive gesture, based on fear and apprehension
Bristling hair down the tail	"I challenge you!"	This tail signal adds an element of threat and aggression to any other tail signal or position
Tail bristling only at the tip	"I'm a little stressed right now."	This tail signal adds an element of fear or anxiety to any other tail signal or position
A crick or sharp bend in the tail	"If I have to, I'll show you who is boss around here!"	This signal adds both dominance and imminent threat to any other tail signal or position
Slight tail wag, each swing of only small size	"You like me don't you?" "I'm here."	A somewhat tentative submissive signal, which can be added to most tail positions
Broad tail wag that doesn't involve the hips or lowered body posture	"I like you." "Let's be friends."	A casual friendly gesture, not involving any social dominance; may also be seen during play

VISUAL SIGNAL	HUMAN LANGUAGE MEANING	CONDITIONS AND/OR EMOTIONS
TAIL SIGNALS (continued)		
Broad tail wag, with wide swings that actually pull the hips from side to side, perhaps with lowered hindquarters	"You are my pack leader and I will follow you anywhere!"	A sign of respect and mild submission for the person or dog to which it is directed. The dog does not feel threatened but acknowledges its lower rank and its confidence that it will be accepted
Slow tail wag with tail at a moderate to low position	"I don't quite understand this." "I'm trying to get the message."	Not really a social signal, but rather a sign of indecision or confusion about what is going on or what is expected of the dog
General Rules about tail signals: *—The higher the tail, the more dominant the signal; the lower the tail, the more submissive.* *—Rate of tail movement indicates degree of arousal and excitement. A tremoring tail (one that seems to vibrate rather than really swing from side to side) should not be interpreted as a tail wag, but is a pure sign of emotion and excitement.* *—All tail signals should be interpreted in relationship to the dog's usual, relaxed tail position (e.g., greyhounds normally carry their tails low, while Malamutes carry their tails high when they are relaxed).*		
BODY LANGUAGE		
Stiff-legged, upright posture, or slow, stiff-legged movement forward	"I'm in charge around here." "Are you challenging me?"	An active aggressive signal from a dominant dog that is willing to assert its leadership
Body slightly sloped forward, feet braced	"I accept your challenge and am ready to fight!"	Usually a response to a threat, or the response to another dog's failure to back down at a threat; a signal that active aggression is imminent
Hair bristles on shoulders and down the back	"I've had it with you! Take your pick: give it up now, fight, or back off!"	A sign of rising aggressive feelings in a confident dominant dog; may indicate that an attack could occur at any moment

VISUAL SIGNAL	HUMAN LANGUAGE MEANING	CONDITIONS AND/OR EMOTIONS
BODY LANGUAGE (continued)		
Hair bristles only on shoulders	"You're making me nervous. Don't push me into a fight." "I don't like this."	Often a sign of fearful aggression from a dog that is threatened but feels it may be forced to fight
Dog lowers its body or cringes, while looking up	"Let's not argue." "I accept that you are higher-status than me."	An active submissive gesture to pacify a more dominant dog
Muzzle-nudge	"You are my leader. Please acknowledge me." "I want . . ."	Much the same meaning as licking behavior, but not quite as submissive; could also be used to ask for things
Dog sits when approached by another, allowing itself to be sniffed	"We're nearly equal in status, so let's be peaceful and civil around each other."	A small pacifying gesture by a usually dominant dog, who is only slightly outranked by another
Dog rolls on side or exposes underside and breaks off eye contact completely	"I'm just a lowly beast that accepts your full authority and am no threat at all."	Passive submission—the dog equivalent of groveling
Standing over another dog who may be lying down. Head over back or shoulders of another dog. Paw over or on another dog.	"I'm bigger, taller, stronger, and really the leader around here."	All of these are mild active assertions of social dominance and social status
Shoulder bump	"I'm dominant over you, and you will give way to me when I come near."	A more vigorous assertion of relative social dominance; a milder version of this same signal is leaning
Dog turns its side toward another animal	"I accept that you are more dominant than me, but I can still take care of myself."	A mild admission of slightly lower social rank by a confident dog, without any fear or distress; if a larger social gap exists, it may turn its hindquarters toward the dominant dog

VISUAL SIGNAL	HUMAN LANGUAGE MEANING	CONDITIONS AND/OR EMOTIONS
BODY LANGUAGE (continued)		
When threatened by another dog: —dog sniffs the ground or digs at something —stares at horizon —scratches itself	"I don't see you threatening me and am not going to respond to it, so calm down."	Pacifying or calming signals based upon distraction; they signal an absence of hostility but no submission
Dog sits with one front paw slightly raised	"I'm a bit anxious, uneasy, and concerned."	A sign of insecurity and mild stress
Dog rolls on its back and rubs its shoulders on the ground (sometimes associated with nose rubbing)	"I'm happy and all is well."	A ritual that often occurs after something pleasant has happened, hence sometimes called a "contentment roll"
Dog crouches with front legs extended; rear body and tail are up	"Let's play!" "Oops! I didn't mean to frighten you. This is all in fun."	Standard play invitation, which may also be used to reassure another dog that rough or threatening behavior was not intended to be taken seriously

General Rules for body language signals:
—Attempts to make the dog look taller or larger represent dominance signals.
—Attempts to make the dog look smaller represent submissive or pacifying signals.
—Pointing body, head, or eyes at another dog represents dominance and perhaps threat.
—Turning body, head, or eyes away is a calming and pacifying signal.

Endnotes

2 Evolution and Animal Language

1. P. Lieberman, *The Biology and Evolution of Language.* Cambridge, MA: Harvard University Press, 1984.

2. J.M. Allman, *Evolving Brains.* New York: Freeman, 1999.

6 The Dog Speaks

1. E.S. Morton & J. Page, *Animal Talk.* New York: Random House, 1992.

7 Learning to Speak

1. J.A.L. Singh & R.M. Zingg, *Wolf-Children and Feral Man.* Hamden, CT: Archon Books, 1966.

2. W.N. Kellogg, *The Ape and the Child: A Study of Environmental Influence upon Early Behavior.* New York: Hafner, 1967.

8 Face Talk

1. P. Eckman, *Telling Lies.* New York: Norton, 1992.

10 Eye Talk

1. G. Bendelow & S.J. Williams, *Emotions in Social Life: Critical Themes and Contemporary Issues.* New York: Routledge, 1998.

15 Signing and Typing

1. R.A. Gardner, B.T. Gardner & T.E. Cantfort, *Teaching Sign Language to Chimpanzees.* Albany: SUNY press, 1989.

2. R. Fouts, S.T. Mills & J. Goodall, *Next of Kin: My Conversations with Chimpanzees.* New York: William Morrow, 1997.

3. D. Premack, *Intelligence in Ape and Man.* Hillsdale, NJ: Erlebaum, 1976.

4. S. Savage-Rumbaugh, *Ape Language: From Conditioned Response to Symbol.* New York: Columbia University Press, 1986.

5. Excerpt from: E.M. Borgese, *The Language Barrier: Beasts and Men.* New York: Holt Rinehart & Winston, 1965 (pp. 64–65).

16 Scent Talk

1. J.P. Scott & J.L. Fuller, *Genetics and the Social Behavior of the Dog.* Chicago: University of Chicago Press, 1965.

2. R. Peters, *Dance of the Wolves.* New York: McGraw-Hill, 1985.

18 Doggish Dialects

1. L.N. Trut, "Early canid domestication: The farm-fox experiment." *American Scientist* (1999) 87; 160–69.

2. D. Goodwin, J.W.S. Bradshaw & S.M. Wickens, "Paedomorphosis affects agonistic visual signals of domestic dogs." *Animal Behaviour* (1997) 53; 297–304.

19 Is It Language?

1. R.I.M. Dunbar, *Grooming, Gossip and the Evolution of Language.* London: Faber and Faber, 1996.

20 Talking Doggish and Doggerel

1. K. Hirsh-Pasek & R. Treiman, "Doggerel: Motherese in a new context." *Journal of Child Language* (1982) 9; 229–237.

Index